Solar Textiles

Most photovoltaic (PV) installations utilise heavy conventional glass or polycarbonate panels, and even newly developed thin plastic or metal films for PV cell use may fracture during both construction and application. Textile fabrics, the most widespread flexible materials in everyday use, offer a solution to the need for lightweight, flexible solar PV generators. *Solar Textiles: The Flexible Solution for Solar Power* is about the incorporation and operation of solar cells on textile fabrics. The combination of textile manufacturing and solar PV cell technology opens up further avenues for both the textile and semiconductor industries. Thus, this book reflects the progressively increasing commercial interest in PV cell technology and the versatility that their integration in textiles provides.

- Discusses textiles as electrical substrates
- Explains the photovoltaic effect and associated parameters
- Offers special consideration of solar cells on textiles
- Compares fibres and fabrics and how to implement PV activity on a textile
- Describes manufacturing methods outside of semiconductor technology
- Includes applications open only to textiles

This work is aimed at textile technologists, electronic engineers, solar technologists, civil engineers and designers in building fabrics and architecture.

Textile Institute Professional Publications

Series Editor:
Helen D. Rowe
The Textile Institute

For more information about this series, please visit: www.routledge.com/Textile-Institute-Professional-Publications/book-series/TIPP

Solar Textiles
The Flexible Solution for Solar Power

Robert Mather and John Wilson

CRC Press
Taylor & Francis Group
Boca Raton London New York

CRC Press is an imprint of the
Taylor & Francis Group, an **informa** business

Cover image by Michael Gonyea with permission from Pvilion

First edition published 2023
by CRC Press
6000 Broken Sound Parkway NW, Suite 300, Boca Raton, FL 33487-2742

and by CRC Press
4 Park Square, Milton Park, Abingdon, Oxon, OX14 4RN

CRC Press is an imprint of Taylor & Francis Group, LLC

© 2023 Taylor & Francis Group, LLC

Library of Congress Cataloging-in-Publication Data
Names: Wilson, John I. B., author. | Mather, R. R., author.
Title: Solar textiles : the flexible solution for solar power / John Wilson and Robert Mather.
Description: First edition. | Boca Raton, FL : CRC Press, 2023. |
Series: Textile institute professional publications |
Includes bibliographical references and index.
Identifiers: LCCN 2022023221 (print) | LCCN 2022023222 (ebook) |
ISBN 9780367706050 (hbk) | ISBN 9780367706029 (pbk) | ISBN 9781003147152 (ebk)
Subjects: LCSH: Solar cells—Materials. | Photovoltaic cells—Materials. | Electronic textiles.
Classification: LCC TK2960 .W565 2023 (print) | LCC TK2960 (ebook) |
DDC 621.31/244—dc23/eng/20220725
LC record available at https://lccn.loc.gov/2022023221
LC ebook record available at https://lccn.loc.gov/2022023222

ISBN: 978-0-367-70605-0 (hbk)
ISBN: 978-0-367-70602-9 (pbk)
ISBN: 978-1-003-14715-2 (ebk)

DOI: 10.1201/9781003147152

Typeset in Times
by codeMantra

Contents

Textile Institute Professional Publications

The aim of the *Textile Institute Professional Publications* is to provide support to textile professionals in their work and to help emerging professionals, such as final year or master's students, by providing the information needed to gain a sound understanding of key and emerging topics relating to textile, clothing and footwear technology, textile chemistry, materials science and engineering. The books are written by experienced authors with expertise in the topic and all texts are independently reviewed by textile professionals or textile academics.

The textile industry has a history of being both an innovator and an early adopter of a wide variety of technologies. There are textile businesses of some kind operating in all countries across the world. At any one time, there is an enormous breadth of sophistication in how such companies might function. In some places where the industry serves only its own local market, design, development and production may continue to be based on traditional techniques, but companies that aspire to operate globally find themselves in an intensely competitive environment, some driven by the need to appeal to followers of fast-moving fashion, others by demands for high performance and unprecedented levels of reliability. Textile professionals working within such organisations are subjected to a continued pressing need to introduce new materials and technologies, not only to improve production efficiency and reduce costs but also to enhance the attractiveness and performance of their existing products and to bring new products into being. As a consequence, textile academics and professionals find themselves having to continuously improve their understanding of a wide range of new materials and emerging technologies to keep pace with competitors.

The Textile Institute was formed in 1910 to provide professional support to textile practitioners and academics undertaking research and teaching in the field of textiles. The Institute quickly established itself as the professional body for textiles worldwide and now has individual and corporate members in over 80 countries. The Institute works to provide sources of reliable and up-to-date information to support textile professionals through its research journals, the *Journal of the Textile Institute* (http://www.tandfonline.com/action/journalInformation?show=aimsScope&journalCode=tjti20) and *Textile Progress* (http://www.tandfonline.com/action/journalInformation?show=aimsScope&journalCode=ttpr20), definitive descriptions of textiles and their components through its online publication *Textile Terms and Definitions* (http://www.ttandd.org) and contextual treatments of important topics within the field of textiles in the form of self-contained books such as the *Textile Institute Professional Publications*.

Acknowledgements

We would like to thank John Andrews, who originally implanted in us our interest in solar textile fabrics and Helen Rowe, who suggested to us the idea of writing this book and who has given us great support during its compilation.

We have also received helpful input from Pauline van Dongen, Neil Morrison, Kathleen McDermott, Lisa Macintyre and Alex Nathanson.

We acknowledge too the work contributed by our students: Adel Diyaf, Suzanne Jardine, Helena Lind and Artem Lukianov.

Finally, but certainly not least, we would like to thank our wives for their patience while the book was being written.

Authors

Robert Mather, PhD, became involved with textiles when he joined the staff at the Scottish College of Textiles (SCOT) in Galashiels in 1983. He had previously worked for 10 years at Ciba-Geigy Pigments in Paisley, following two spells of postdoctoral work at Imperial College and Brunel University in London. At SCOT, which subsequently became part of Heriot-Watt University, Dr Mather acquired his interest in technical textiles and established a profile in the processing of polypropylene fibres. He also worked on three-dimensional woven engineering products.

His current R&D interests include the incorporation of solar cells on textiles, and Power Textiles Limited was formed in 2012 with John Wilson as Co-Director to promote solar textiles. Robert also has an interest in gas plasma treatments for textiles. He was Director for 5 years of the newly established Technical Textiles and Polymer Innovation Unit at Galashiels, set up with funds from the UK Department of Trade and Industry. During this period, the Unit successfully undertook projects for a variety of companies, and Robert was contracted to write on the technical textile industry in Scotland for a briefing paper for the Scottish Executive. The Unit was eventually subsumed into the UK TechniTex Faraday Partnership, and he became responsible for technology transfer in the Partnership's organisation. He has also been Director of the Biomedical Textiles Research Centre at Heriot-Watt University.

He is the author of over 90 technical papers and in recent years has contributed chapters to a number of books on technical textiles. Together with a co-author, he wrote a book on textile chemistry for the Royal Society of Chemistry in 2011. A second, updated edition was published in 2015, and a third edition is currently in preparation.

His professional memberships include Fellow of the Royal Society of Chemistry, CChem, and Fellow of the Textile Institute, CTex.

John Wilson, PhD, was Professor of Materials Processing at Heriot-Watt University, Physics, from 1996 to 2012. His first post there was Wolfson Research Fellow 1975, following a similar appointment at St Andrews University. At Heriot-Watt University, John was Academic Head of Physics (2002–2005 and 2010–2012), setting up several degree programmes and supervising over 30 PhD and MSc students, including a collaboration with Chinese PhD students from Beijing Jiaotong University. His publications include a book on solar energy, encyclopaedia and book chapters and over 200 scientific papers.

Dr Wilson's interests in the applications of solar energy span over 40 years of research, from thin-film II–VI cells for his PhD thesis, later pioneering the UK amorphous silicon solar cell research, and to current work on flexible solar cells. Through Rotary, he worked with a small group at The Turing Trust to design and deliver a solar-powered, off-grid, computer laboratory for Malawi. He has been a consultant to the UN and to UK companies and institutions throughout the world (e.g. Romania, India, Nigeria) on solar energy and photovoltaics, and was founder secretary of the

Scottish Solar Energy Group until 1988. He presently delivers solar photovoltaic training courses for the Renewable Energy Institute Ltd, Edinburgh.

His materials R&D used lasers and plasmas to deposit or modify metals, semiconductors, dielectrics and polymers, for applications in optics, mechanics and bioengineering. In particular, research into thin-film diamond led to the founding of DILAB Limited in 1994, of which he was managing director. Power Textiles Limited was founded in 2012 with Co-director Robert Mather to promote flexible solar cells on textiles.

His professional memberships include Fellow of the Institute of Physics (and a past member of their Science Board), CPhys and CEng.

1 The Versatility of Textile Fabrics

The juxtaposition of the words 'solar' and 'textiles' may not at first sight seem very evident. The term 'solar' in this context refers to the photovoltaic (PV) effect, whereby light is directly converted to electricity. The PV effect is the basis of operation of solar panels, commonly observed for example on household and office roofs and arrayed in fields in the countryside. On the other hand, textiles are mainly thought of in terms of clothing, fashion and household furnishings, although there are additionally many highly technical uses of textiles such as in construction, agriculture, engineering, military and medicine. It is clear then that textiles enjoy a wide range of applications. The increasing interest in electronic textiles (products that contain both electronic and textile components in a single device) opens paths to still more applications, and in this book we set out to highlight one of these: the potential of textiles as flexible supports for solar cells. Hence, we arrive at solar textiles!

Solar energy is just one of several sustainable energy sources. Other sources include wind turbines, wave power, tidal power and hydroelectric schemes. The one source, however, that will provide an 'endless' supply of energy is that fusion reactor in the sky we call the sun. It is a well-publicised fact nowadays that more energy is provided by the sun to the whole world in 1 hour than the energy used in 1 year by the world's population. Moreover, PV cells provide a very efficient method of directly harnessing solar radiation. The environmental benefits of PV cells are considerable. During their operational lifetime, generally around 30 years, they emit no greenhouse gases into the atmosphere. Moreover, there are no moving parts that require regular servicing or replacement. Hence, maintenance costs are very small, and so the energy input the solar cells require for operation during their lifetime is very low indeed. Input of energy is, however, required for the manufacture, transport, installation and decommissioning of solar cells, and, as discussed in Chapter 7, estimates of this input vary greatly for any particular type of cell. Nevertheless, even the most pessimistic of these estimates indicates that after only 3–4 years of operation the total energy generated by the cell will have overtaken the amount of energy input required. The energy generated during the remaining years of a cell's operational life is 'free'.

The advantages offered by textiles as vehicles for solar cells are covered later in this chapter, but we may note here that this development has also been partially driven by the miniaturisation of electronic devices during the past 20–30 years and the consequent capability to integrate active and passive circuitry together with power sources such as solar cells on textiles. Considerable progress has been made in the incorporation of electronic sensors into wearable fabrics although, most notably in clothing, functionality of the sensors has had to be balanced with the comfort and aesthetics of textiles. There have been particular successes in sports and medical

DOI: 10.1201/9781003147152-1

applications where, respectively, sensors in sports tops and sports bras can monitor the heart rates of athletes and sensors in specially adapted clothing can check heart function in younger babies and the elderly. These sensors do, however, have to be powered, either directly from external sources or via small battery intermediates.

The integration of solar cells into the fabric offers a route to providing the power required. In daylight or under artificial light, a solar cell can directly power a sensor, with both embedded in the fabric and electrically connected to each other. In darkness, of course, the cell cannot function. For this reason, some or all of the energy generated by the solar cell needs to be stored in batteries (or perhaps supercapacitors, as outlined in Chapter 9), so that the sensor can still be powered in the absence of light. It would be desirable then if the batteries could also be embedded in the fabric. The ongoing developments of flexible thin-film batteries and printed batteries are now making this goal a realistic one. Thus, a textile fabric could incorporate integrated solar cells connected to embedded batteries that are, in turn, connected to the sensors to be powered.

There are a number of approaches to adapting textiles as supports for solar cells. The simplest approach is to attach small solar cells to a fabric; but whilst the performance of the cells will be maintained, the properties of the fabric will be considerably compromised. Two alternative strategies that would produce genuinely integrated solar textiles are the deposition of cells on yarns that are subsequently fashioned into a fabric and the deposition of cells onto a finished fabric. Both these strategies require the initial deposition of electrical contacts and semiconducting layers onto substrates possessing quite different characteristics from those used in conventional PV technology.

The solar panels that are commonly recognised consist of interconnected solar cells applied to glass, or maybe polycarbonate plates (Figure 1.1). Solar panels have achieved substantial success over a number of years – and continue to do so – but they nonetheless have a number of drawbacks. Most obviously, they are flat and inflexible and so can be attached only to flat surfaces. Moreover, they are heavy, so

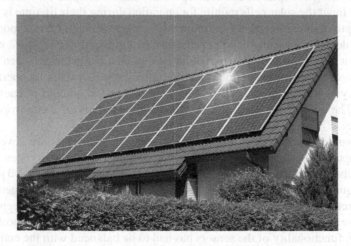

FIGURE 1.1   Conventional solar panels on a house rooftop. (Shutterstock.)

FIGURE 1.2 Flexible 7 watt solar panel of dimensions 30 cm × 70 cm.

the structures to which the panels are to be attached must be strong enough to accommodate them. Also, they are fragile. Their storage, transport and installation therefore require some care. The use instead of a light-flexible support has been seen as a means of circumventing these problems. To this end, a wide range of PV cells is now available commercially that are applied to thin plastic or metal films (Figure 1.2). The products can maintain their durability in quite challenging environments. Other benefits are lower materials use and reduced construction costs.

Nevertheless, the thin nature of these films may well render them prone to fracture during construction and installation. A more suitable alternative then would be an already existing type of flexible substrate that can stand up to the conditions required for the deposition of PV cells, their installation and their subsequent use. Textile fabrics, therefore, would be an attractive option. They are of course very familiar to everybody and have been for thousands of years. They are still the most widely used types of flexible material and, as already noted, are put to a very wide range of applications. There are therefore sound arguments for developing their use as flexible supports for PV cells.

In practice, the integration of solar cells on textiles is technologically challenging. One solution that we shall meet quite often in this book is the adhesion of thin PV films to textiles. However, the mechanical properties of these thin films and those of textiles are far from identical. As already noted, such thin films are liable to rupture, especially in a harsh environment. In addition, there can be a real risk of the film separating from a textile yarn or fabric, particularly where the yarns are continually being flexed and then straightened, as when clothing is worn. Strategies to overcome these difficulties are discussed in Chapter 5.

Nevertheless, despite the technological challenges involved, the goal of producing truly integrated PV textile fabrics is certainly one worth striving for. Indeed, Chapter 8 surveys a whole host of commercial examples where solar textile fabrics are already being applied. Even some aesthetic applications are included.

In this book, we explain how PV cells and textile fabrics are brought together to produce solar textiles. Chapter 2 explains the PV effect, the construction of PV cells in general and some of the advances in materials and cell configuration that take us beyond conventional silicon cells into thin-film cells. Chapter 3 explains why the actual construction of a textile fabric is so important for the deposition and function of an array of PV cells. The many different types of fabric construction are surveyed, and the merits and drawbacks of each type as vehicles for PV cells are discussed. Chapter 4 explains the need for textile fabrics to be rendered electrically conducting before the deposition of PV cells. The numerous means of achieving electrical conductivity are surveyed. In Chapter 5, we discuss how to render textile fabrics PV. Three alternative approaches are described and critically analysed: attachment of individual cells or PV films to a fabric, fabric construction from yarns already rendered PV and the direct deposition of PV cells on a fabric.

Chapter 6 provides an overview and a critique of technological and design specifications for PV textile fabrics. Special aspects that relate to PV textile devices, as distinct from traditional photovoltaics, are highlighted. Chapter 7 covers the transition from laboratory-scale fabrication to pilot production and beyond. The environmental impact of using textile substrates in place of traditional ones is analysed. In Chapter 8, we highlight some of the many applications open to commercially produced PV textile fabrics. The opportunities and limitations for these fabrics in the fashion world are assessed, and a number of commercial sectors are discussed where the integration of PV cells can provide additional benefits to a product.

Finally, in Chapter 9, we consider the outlook for solar textile fabrics. Conventional PV arrays will in practice be very seldom sufficient on their own but will need to be integrated with energy storage devices that are not immediately fed to the item to be powered. The state-of-the-art and future developments in flexible batteries and supercapacitors are discussed, and the future of flexible combinations in textiles of PV devices, electrical storage devices and sensors is considered.

Solar-powered textiles will clearly have a significant part to play in the sustainable energy story. Indeed, Chapter 8 demonstrates that there is already wide-ranging interest in them now. We hope readers enjoy their journey through this book.

2 The Photovoltaic Effect and How It Is Used

2.1 INTRODUCTION

Solar cells use the photovoltaic effect to convert light directly into electricity and the underlying principles are the same for all types of cells, regardless of materials or construction. Because the effect involves neither chemical reactions nor movement, and does not require heat exchange, there are no causes of degradation to reduce the operating lifetime unlike batteries that employ electrochemical processes. In order to understand why this is so, some consideration of solid-state physics is essential, and this will provide a better basis for explaining the fundamental limitations on the efficiency of energy conversion which in any case can never be 100%. In this regard, the thermodynamic rules for explaining the operation of heat engines apply in a similar fashion to photovoltaic (*PV*) cells, with the temperature of the heat source being replaced by the intensity of the illumination.

The following sections will build from a simple explanation of the photovoltaic effect to a more detailed version, followed by a description of PV cell construction and the materials that are currently being used. This will lead to the specific types of cells that may be made in thin-film and flexible forms and will be concluded by a brief examination of newer materials that offer improvements in performance, albeit still limited by the basic laws of energy conversion.

2.2 LIGHT INTO ELECTRICITY

The technique for achieving this transformation trailed behind the use of solar energy for heating as it required the advent of semiconductors. Three hundred years ago there were many wild schemes for making money based on pseudoscientific ideas, which were satirised in Dean Swift's 'Gulliver's Travels to Laputa' written in 1726. This describes a visit by Gulliver to the fictional academy of Lagado, where he meets a professor engaged in a project to extract sunbeams from cucumbers, to be sealed into phials and let out to warm the air in raw inclement summers. Nowadays we understand that energy may be converted from one form into another, although not without some losses: for instance we may convert mechanical stress directly into electric charge by the piezoelectric effect, or differences in temperature directly into voltage by the thermoelectric effect. Cucumbers could well be considered as a store of solar energy but converting this natural process back into sunbeams is indeed not possible!

What *is* possible is a variety of photo-effects that involve electricity: photoemissive, photoconductive and photovoltaic effects. The first two of these require an additional source of electric potential, i.e. voltage, but the last one uniquely generates

DOI: 10.1201/9781003147152-2

PHOTOVOLTAIC EFFECT

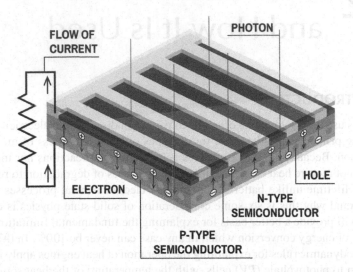

FIGURE 2.1 Schematic diagram of a PN junction solar cell showing the photogeneration of positive and negative charges that then flow in opposite directions across the junction. (Source: Shutterstock.)

both voltage and current to deliver power. Photoemissive sources release electrons from an illuminated surface into a vacuum, known as the photoelectric effect; photoconductive devices change their electrical conductivity according to the light they absorb.

Simply put, photovoltaic devices work by absorbing light to generate pairs of positive and negative charges, which are separated by a built-in voltage to provide a current at the output terminals. The key to generating useful power is having a built-in voltage, which is provided by two semiconducting materials in close contact. The choice of semiconductor determines the colour of light that may be absorbed, and the built-in voltage arises from the junction between positive-type (*p-type*) and negative-type (*n-type*) semiconductor variants, with the addition of conducting contacts on each to complete the cell. Most deployed photovoltaic solar cells use crystalline silicon wafers with an opaque metal layer on the back and a metal grid on the front to allow sunlight into the silicon (Figure 2.1).

2.3 THE PHOTOVOLTAIC EFFECT

The apparent simplicity of the device masks the manufacturing precision needed to refine metallurgical-grade silicon for electronic requirements, as well as the practical cell features that enable manufacturers to provide the best performance. There has been a history of continuous enhancement since the first silicon photovoltaic cells were developed in 1954 at Bell Laboratories in the USA by Chapman, Fuller

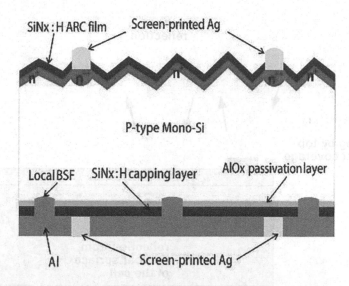

SiNx:H ARC film   Screen-printed Ag

P-type Mono-Si

Local BSF   SiNx:H capping layer   AlOx passivation layer

Al   Screen-printed Ag

FIGURE 2.2 Cross-section of a high-efficiency silicon solar cell (PERC type: passivated emitter and rear cell). The textured front is coated with an antireflection film (ARC), and the rear has a thin dielectric layer with fine holes that enable electrical contact with the back surface field region (BSF). (From Bai, Q., Yang, H., Cheng, X. and Wang, Recombination Parameters of the Diffusion Region and Depletion Region for Crystalline Silicon Solar Cells under Different Injection Levels. *Appl. Sci.* 2020, 10, 4887.)

and Pearson, moving from a few percent power conversion efficiency to ~25% at the present day. These performance enhancements come with increased costs but have been responses by manufacturers to intense competition in the industry. This section will show where the conversion losses occur and how they have been addressed by structural improvements (Figure 2.2).

A comparison between the early cells and the advanced cells available today reveals the changes from a simple planar configuration to a more structured arrangement. Textile users are already familiar with a complex surface made from yarns or filaments, but the microelectronics industry was founded on planar geometry that required very smooth surfaces for alignment of succeeding layers. This characteristic of fabrics may be a challenge for photovoltaic production engineers but could also be an advantage for collecting as much light as possible, as we shall see later.

Following the journey taken by incident light through a typical cell shows how alternative routes arise to the desired one of producing electrical charges. In any sequence of steps, it is the first that has the most influence on the end result; thus it is important to transmit as much incident light as possible through the cell surface. Some light will be blocked unavoidably by the conducting grid contact. Reflection losses at this first face may be reduced by antireflective coatings or by careful surface roughening. Light passing into the active absorbing material, most commonly silicon, will only be absorbed if it has the correct colour (Figure 2.3).

This is a consequence of quantum physics which is that light must have sufficient energy (as shown by its wavelength or colour) to release electrical charges from

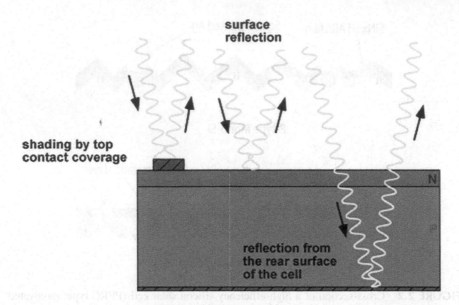

FIGURE 2.3 Paths taken by light beams incident on a solar cell showing reflection and shadowing losses. (From C.B. Honsberg and S.G. Bowden, "Photovoltaics Education Website," www.pveducation.org, 2019.)

their bindings to silicon atoms, but not so much that the minimum energy is greatly exceeded. This restriction means that only a band of the complete light spectrum may be absorbed in any particular semiconductor, which sets us one parameter for selecting the best material. Light of the correct wavelength for absorption may still be partially transmitted through a semiconductor if it is too thin, but if it is too thick then the extra material is a wasted expense. However, light that has not been absorbed will reach the rear face of the cell and may be reflected back for a second pass.

The light that is usefully absorbed will generate mobile electrical charges, one pair of positive and negative charges for each absorbed photon, the fundamental quantum of light that describes its energy, depending on the colour. (Although the optical path may be described by geometry using the wavelength of light, the conversion of light into free charges is described by quantum mechanics using photons.) These free charges will recombine and release the energy taken from the photon, unless they are quickly separated. This requires an electric field which is provided by the interface between p-type and n-type semiconductors.

We cannot measure this field directly but it is produced by fixed charges on specially added impurities, known as dopants, on each side of the junction. Dopants are introduced in parts per million so do not disturb the chemical composition of the pure semiconductor. They are most active when they can replace a host atom without disturbing the regular crystal structure of the host semiconductor. We can incorporate boron into silicon to make it p-type and phosphorus to make it n-type. The more that is added, the stronger the electric field, but also the thinner the region over which it operates. This is another trade-off that device designers must optimise.

Energy is also lost whenever charges recombine before they are delivered to the cell's electrical contacts, negative charges (*electrons*) go to the *negative-type* or n-type side and the positive charges (*holes*) go to the *positive-type* or p-type side. Imperfections or chemical impurities in the crystal structure are sites for charges to collect and recombine, as are surfaces unless additional electric fields are included. High-energy photons, those towards the blue end of the spectrum, have more energy than is needed to release electrons from their bindings, and the excess above this minimum is lost as heat; lower-energy photons that can still be absorbed may travel further into the semiconductor away from the built-in field and so generate charges that are not separated. Some semiconductors have an inherent property of producing pairs of charges that remain closely connected (*excitons*) and require well-designed fields to separate and collect these.

Taken together, these optical and electrical losses lead to a maximum conversion efficiency for sunlight into electricity of between 25% and 30% according to the detailed spectrum used in the calculation. However, this is not very much lower than thermal conversion of steam to electricity in conventional power stations. Photovoltaic efficiency may be increased beyond this theoretical upper bound by having multiple junctions per cell with two or more different semiconductors, to enable more than a single band to be absorbed from the solar spectrum. This comes with the price of increased manufacturing complexity. Greater efficiency is also possible by using concentrated sunlight from mirrors or lenses, akin to increasing the temperature of steam in conventional electricity generation. These additions have attained efficiencies greater than 47%.

2.4 SOLAR CELL PARAMETERS

So how do we characterise the performance of a solar cell? There are some parameters that must be provided so that designers can estimate the output power for a particular situation. There has been international agreement on the standard illumination conditions under which terrestrial solar cells must be tested, known as the standard test conditions (*stc*). (These do not apply to cells intended for indoor illumination.) The stc values are an intensity of 1 kW/m², an AM1.5 solar spectrum and a temperature of 25°C. The solar spectrum outside the atmosphere is at AM0, and the air mass 'AM' number is the equivalent thickness of the earth's atmosphere that the sun's radiation passes through to reach the earth's surface at any location. Thus when the sun is directly overhead, the air mass is 1.0, increasing as the sun lies lower in the sky. The stc are not found in actual outdoor conditions but are realisable with solar simulator lamps for test purposes.

The solar cell or solar panel output should be given as the maximum power conversion efficiency under stc, together with the current and voltage at this value. The maximum power point P_{max} lies between the short-circuit current I_{sc} and open-circuit voltage V_{oc} points on the characteristic current versus voltage ($I \sim V$) curve. These are the extreme values delivered by the cell or panel when its contacts are either short circuited (i.e. having zero output voltage) or disconnected from any load (i.e. having zero output current). The significance of the maximum output power will be seen later. Another parameter that may be met is the Fill Factor (FF) which is the ratio

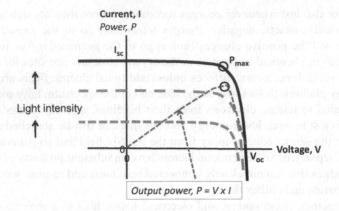

FIGURE 2.4 Current versus voltage characteristics for a typical solar cell showing the short circuit current, open circuit voltage and maximum power points for the curve receiving the highest illumination. Also shown is the power versus voltage curve.

of the maximum power to the product of I_{sc} and V_{oc}, effectively a measure of how 'square' the curve is, which has a value of over 80% in high-efficiency conventional solar panels (Figure 2.4).

As implied above, the output of a solar cell will depend on its illuminating source, both intensity and spectrum. Quite simply, the output current close to the short-circuit condition will directly depend on the intensity of the light, whereas the output voltage will change only slowly with intensity. A further ambient condition that greatly affects performance is the temperature: the higher the temperature, the lower the output power, especially at the open-circuit voltage end of the $I \sim V$ curve. This is unfortunately true for most types of solar cells.

2.5   SOLAR CELL MATERIALS

The electrons that are freed by illuminating a silicon cell are those that are least strongly bound to the silicon atoms, and in chemical terms are responsible for holding the atoms together in a crystalline lattice: they are the *valence* electrons. In semiconductor science, they are said to be transferred from a *valence band* of energy to a higher *conduction band*. The binding energy of these electrons is identical to the energy separation of valence and conduction bands, known as the *bandgap*. Each semiconductor has its own specific bandgap, setting the minimum energy that must be absorbed to generate conduction electrons.

The ideal bandgap for absorbing solar energy lies in the visible part of the spectrum not far from the bandgap of silicon. This is fortuitous because silicon microelectronic technology is the most advanced of any semiconductor. Compound materials such as gallium arsenide (GaAs) and cadmium telluride (CdTe) are better matches to the solar spectrum but are more difficult to process and more costly. (GaAs is used for applications where efficiency rather than cost is the principal issue, and CdTe is one of the materials that may be fabricated as an effective thin-film absorber, thus lowering the cost.)

Semiconducting silicon is a crystalline material, manufactured as rather brittle thin wafers, only 50–75 mm in diameter when solar cells first became a commodity, but now often 182 mm for the best performing panels. Their production route from silicaceous sand to electronic-grade wafers is demanding in both energy and technique. This sets a base cost which cannot be easily reduced. One route to avoiding this is to use a thin-film coating of semiconductor on a cheaper substrate such as glass. Silicon is virtually impossible to produce as a crystalline thin film as it oxidises so readily, but alloys of cadmium and tellurium – or of copper with indium, gallium and selenium (CIGS) – are more amenable, despite their more complex composition. They are both used in solar panels that today reach similar efficiencies to the simpler forms of silicon cell.

One form of thin-film silicon that is used for low-cost solar cells is amorphous silicon, actually a non-crystalline alloy of silicon and hydrogen, a-Si:H. This is a more effective semiconductor than pure thin-film silicon, with a bigger bandgap (closer to the optimum for solar energy conversion), but is less perfect as an electronic conductor and is only ~10% efficient. However, it is made by a lower temperature process than for silicon wafers and so is compatible with a range of substrate materials. The inclusion of hydrogen passivates *dangling* bonds that cannot join up in the random arrangement of silicon atoms but also leads to degradation in the most efficient cells under strong illumination.

This lower temperature processing is shared by organic semiconductors, either using small organic molecules or more widely, polymers. Despite this significant advantage, organic PV cells are not widely available as they are sensitive to both oxygen and moisture and have a lower efficiency than competing materials. Nonetheless, they are the preferred material of many research and development teams working with flexible solar cells. Their low efficiency derives from the poor separation of photo-generated charges by the built-in field: electrons and holes are not fully separated when created and remain loosely bound together as an *exciton* pair. The formation of an effective separating field is the aim of the more complicated structures that are used by the best organic cells (Figure 2.5).

Even more complex are the dye-sensitized solar cells (DSSCs) that have been under development since the late 1980s, from the first device by Michael Grätzel after whom they are sometimes named. Light is absorbed in a film of dye on a transparent support such as titanium dioxide, producing excited electrons that are injected into the n-type TiO_2 and oxidising the dye. The electrons then flow via a transparent conducting oxide electrode (typically ITO) through an external electrical load and then on to a metal counter-electrode. The circuit is completed by an electrolyte (an organic solution of iodide ions in the earliest cells) through which the electrons return to the dye, reducing it again. Electron transport through the electrolyte (e.g. iodine in the early cells) is by a reversible redox reaction (e.g. I_3^- to $3I^-$, and back again at the dye surface). (In effect, positively charged holes are injected into the electrolyte at the dye surface.) The participation of a chemical reaction, even if reversible, can lead to ageing that gradually reduces the cell efficiency. Dye must also be chosen to resist photo-bleaching. Despite the need for a sealed package to contact the liquid, these

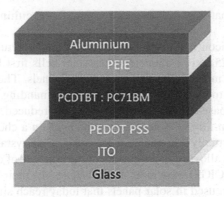

FIGURE 2.5 A typical organic solar cell comprising PEDOT:PSS as the hole transport layer, polyethylenimine ethoxylated (PEIE) as the electron transport layer and a polymer active layer of PCDTBT:PC71BM. (From Lechene, B.L., Cowell, M., Pierre, A., Evans, J.W., Wright, P.K. and Arias, A.C., Organic solar cells and fully printed super-capacitors optimized for indoor light energy harvesting, *Nano Energy*, 2016, 631–640.)

devices are relatively simple to construct in comparison to solid-state semiconductor cells. The solid-state variant replaces the electrolyte with an organic p-type semiconductor that acts as a hole transport layer (Figure 2.6).

In recent years, another class of photovoltaic material, perovskite, has led to much excitement. These have the same crystalline structure as the calcium titanium oxide mineral for which they are named but are composed of a mixed organic-inorganic hybrid such as methylammonium lead iodide. They rapidly attained promising efficiencies, and today have exceeded 25% in development cells. Even higher efficiency is possible by forming a tandem device with crystalline silicon and so this pair is under intense development for the market. Thin-film perovskite solar cells have been made with different structures, and there are great efforts to replace lead with another element that is non-toxic, whilst retaining the efficiency of the lead material. Effective encapsulation is also a necessity as they share the organic PV sensitivity to moisture.

2.6 CONSTRAINTS ON CHOICES FOR FLEXIBLE PV

We can see that the demands of a flexible solar cell or panel eliminate the standard widely used semiconductor materials, including silicon, but still offer some choice. Any thin-film coating must use a low-temperature technique whether this is electrical plasma based (a-Si:H) or liquid based (polymers). Post-deposition processes that are essential for forming the junction must also be low temperature, and this rules out some CIGS-type structures. The electrical contacts must also use a low-temperature process, and we shall see later that there are several options to manufacture these. Finally, note that the standard test conditions that are mandatory for rigid solar PV panels may not apply to flexible solar PV devices that are intended for non-solar illumination, although additional tests will be expected to confirm their stability when flexed.

(a)

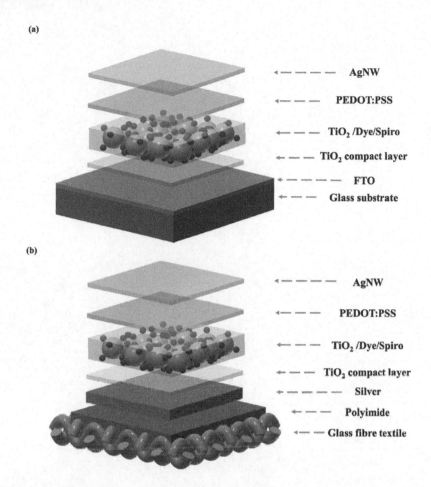

(b)

FIGURE 2.6 Solid-state dye-sensitized solar cell (DSSC) on a glass substrate and a similar construction on a woven glass fibre textile, both using the solid electrolyte, spiro-OMeTAD. (From Liu, J., Li, Y., Yong, S., Arumugam, S. and Beeby, S., Flexible Printed Monolithic-Structured Solid-State Dye Sensitized Solar Cells on Woven Glass Fibre Textile for Wearable Energy Harvesting Applications. *Sci Reports*, 2019, 9, 1362.)

3 Constructions of Textile Fabrics

3.1 INTRODUCTION

The advantages of flexible supports for solar cells and, even more, the merits of textile fabrics as the flexible supports of choice have already been argued in Chapter 1. However, unlike other supports for solar cells such as glass and thin films, textile fabrics are inhomogeneous and display directional properties. In addition, they are highly porous: they typically contain 60%–90% void volume. Some of the porosity arises from voids in the yarns constituting the fabric and some from the actual construction of the fabric. On the other hand, in opposition to these possible drawbacks, textile fabrics are unique in the multitude of constructions that can be adopted through variations in woven and knitted design, and in the hierarchical nature of these constructions. In addition, there is enormous scope for tailoring required shapes for particular constructions with a very wide range of properties, applications and markets. Nevertheless, some constructions are going to provide more effective platforms than others for the deposition and function of a PV array.

After a PV array has been integrated onto a textile fabric, the fabric should retain as far as possible all its original properties, yet enhance PV performance over similar cells that have been deposited on a smooth, homogeneous substrate. There are, however, major differences between the mechanical properties of textile fabrics and those of PV cells. As shown in Chapter 2, the PV array will consist of thin electrically conductive layers and thin semiconductor layers. Even though these layers will be very thin (of the order of micrometres), they will still be inherently rigid and quite brittle, yet they must withstand stretching, bending and twisting of the textile fabric. As a consequence, there are risks of cracking and delamination of the layers. The activity of the PV cells will be impaired, and the conductivities of the conducting layers will be severely reduced. In addition, delamination is more likely to occur if there is little chemical compatibility between the yarn surfaces and the layers adjacent to these surfaces. The requirement for continuity in the deposited layers thus places demands on the dimensional stability of the fabric structure, and in turn the textile fabric can make heavy demands on the durability of the deposited layers. In practice, therefore, there have to be compromises in PV cell design.

In addition, the texture of fabrics makes the continuity of these layers more difficult to achieve compared with the surfaces of smooth substrates, yet the roughness of a textile surface may provide increased optical absorption for shallow angles of incidence of incoming light. Where individual yarns have been coated before fabric construction, successful permanent connections between the many minute PV cells in these yarns become more challenging. Finally, depending on the nature of the semiconducting layers to be deposited, the fabric may need to withstand high processing

DOI: 10.1201/9781003147152-3

temperatures, especially in the case of inorganic semiconductors. The fabric therefore needs to be resistant to melting and degradation up to these temperatures.

We thus need to consider the composition and construction of textile fabrics, in
order to gain an insight into their effectiveness as PV supports during fabric lifetimes
and their ability to withstand the conditions under which photovoltaic PV cells are
deposited, be it as yarns prior to fabric construction or as finished fabrics. Important
aspects of the nature of the fabric are going to include:

 the dimensional stability of the fabric structure;
 the contours of the fabric surface;
 the effects of fabric flexibility and drape;
 the porosity of the fabric;
 the resistance of the fabric to abrasion;
 the chemical nature of the constituent fibres;
 the resistance of the textile to the conditions for depositing photovoltaic arrays;
 the extent to which yarns touch at crossover points in the fabric and the distor-
   tions in the yarns at these points.

The relative importance of each of these factors will be dependent not only on the
constitution and construction of the fabric but also on the application to which the PV
fabric is ultimately assigned. Many of the specifications required for a flexible clothing fabric, for example, will be different from those required of a robust tent fabric.

3.2 PROPERTIES OF TEXTILE FIBRES

The basic building blocks of a fabric are of course textile fibres, and amongst fibres,
there are many physical and chemical differences. Nevertheless, fibres possess one
common characteristic – their fineness: they are long and thin. Numerous fibrous
structures exist in nature, though only those fibres that can be converted to yarns are
suitable for the construction of textile fabrics. Nearly all natural textile fibres, such
as cotton and wool, exist as so-called 'staple' fibres, whose length is generally in the
range of 2–50 cm but whose cross-section is only 10–40 μm. Staple fibres are twisted
together in a spinning process to form yarn. The individual fibres need to possess
some degree of surface roughness so that they adhere to one another in the yarn. By
contrast, synthetic fibres are produced as thin continuous filaments, which can be
twisted together to form a multifilament yarn. However, for some applications, notably knitwear and carpets, these yarns may be cut for conversion into staple fibres of
a desired length. When these staple fibres are subsequently spun, the yarn produced
possesses a softer, bulkier character. Synthetic fibres are also produced as single
monofilaments for more robust applications. It is worth noting here too that, unusu-
ally amongst natural fibres, silk is also produced in the form of continuous filaments
by silk worms and spiders.

Another important feature influencing fabric construction is fibre morphology:
the physical nature of an individual fibre. The morphologies of natural fibres are
complex. For example, cotton fibres are shaped like ribbons, with many convolutions
along their length (ca. 10 per mm). There are four main parts to the fibre structure,

as illustrated in Figure 3.1. The outer layer is a waxy cuticle, just a few molecules thick, that is largely removed during the bleaching and scouring processes of cotton fabric production. Underneath the outer layer is a primary wall consisting of very fine fibrils of cellulose chains. The secondary wall comprises the bulk of the fibre and consists essentially of three layers of cellulose fibrils, spiralling along the fibre axis. The layers are denoted by S_1, S_2 and S_3 in Figure 3.1. The lumen is a hollow that runs along the centre of the fibre.

Wool fibres possess particularly complex structures that include overlapping scales along the surface, with the result that the friction in the root-to-tip direction is much greater than in the opposing direction. Figure 3.2 shows some examples. This so-called directional frictional effect underlies the innate ability of wool to felt, involving the progressive entanglement of wool fibres when they are subjected to

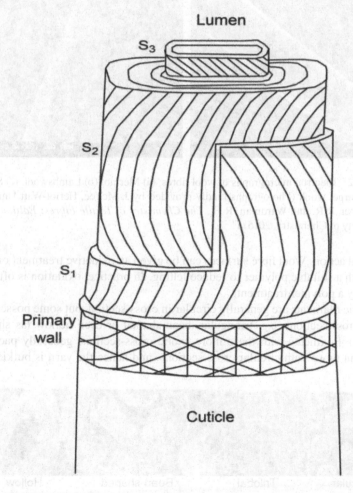

FIGURE 3.1 Schematic illustration of the morphology of a cotton fibre. (From Mather, R.R. and Wardman, R.H., *The Chemistry of Textile Fibres: Edition 2* (The Royal Society of Chemistry, 2015).)

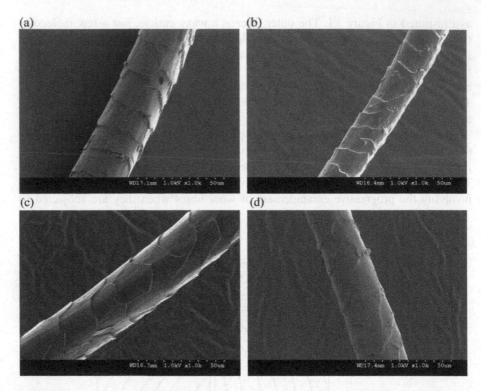

FIGURE 3.2   Electron micrographs of wool fibres. (a) Merino. (b) Lambswool. (c) Shetland wool. (d) Carpet wool. (Photographs kindly provided by J. McVee, Heriot-Watt University.) (From Mather, R.R. and Wardman, R.H., *The Chemistry of Textile Fibres: Edition 2*; (The Royal Society of Chemistry, 2015).)

mechanical action. Wool fibre surfaces can be given an oxidative treatment or can be treated with a suitable polymer to reduce felting. In practice, oxidation is often then followed by a polymer treatment.

Synthetic filaments are generally circular in cross-section, but some possess more complex cross-sections, e.g. triangular, bean-shaped or even hollow (as shown in Figure 3.3). Filaments with these more exotic cross-sections generally pack more loosely than those with circular cross-sections, and hence the yarn is bulkier. Tape

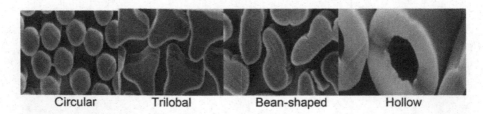

| Circular | Trilobal | Bean-shaped | Hollow |

FIGURE 3.3   Cross-sections of some synthetic filaments. (Photographs courtesy of BISFA, Bureau International pour la Standardisation des Fibres Artificielles.)

filaments, with rectangular cross-sections, are also produced for some applications. Finally, we should include regenerated fibres, whose constituent polymers are extracted from natural sources. Examples include cellulose-based fibres such as viscose and lyocell fibres, both derived from wood pulp, and protein-based fibres such as those derived from soybean and milk casein.

The chemical nature of the polymers comprising textile fibres also plays a part in the properties of the resulting fabrics. One obvious consideration is the ability of the polymeric chains to withstand prolonged exposure to light, especially ultraviolet radiation. For example, polyamide (nylon) and polypropylene fibres tend to degrade on prolonged exposure to light, unless stabilisers have been added during their manufacture that confer resistance to ultraviolet radiation. Polyester (PET) fibres are more resistant to light. The polymer must also be able to withstand the elevated temperatures that are often required to deposit the thin PV layers, particularly for established inorganic semiconductors. The lowest temperature that can be achieved is for the deposition of amorphous silicon at ca. 200°C. Even so, the choice of commodity textiles that can be used is still limited. Polypropylene melts below this temperature. Wool, cotton, silk and acrylic fibres start to decompose below this temperature. PET fibres, however, melt at 260°C–270°C. Thus, for most commodity textiles, deposition methods that do not require elevated temperatures, such as printing, will be preferable, though these technologies are less advanced.

Many high-performance textiles could be used as substrates, because they are stable at temperatures up to 300°C–400°C. Examples include polyimide (PI), polyetheretherketone (PEEK) and polybenzimidazole (PBI). Aramids, such as Kevlar® and Twaron®, would themselves be insufficiently stable to ultraviolet radiation (Mather & Wardman, 2015), although ultraviolet stabilisers are often incorporated into them. However, these textiles are expensive, so there are clear commercial attractions to adopting PET or other suitable commodity textiles. Another option may be fabrics made from E-glass fibres since E-glass is competitive with polyester in price. By far the majority of glass fibres are made from E-glass, which is a calcium alumino-borosilicate glass. The transparency of E-glass could well be advantageous, as with plate glass in conventional solar panels. E-glass, however, has inferior resistance to acid and alkalis, and E-glass fibres are prone to rupture when flexed. Another variant is S-glass, which contains no calcium but has an appreciable magnesium content. S-glass has been used for aerospace components and specialist sports equipment but is much more expensive than E-glass (Jones, 2001).

A further consideration will be the nature of the fibre surfaces in a fabric, as this will influence the performance and adhesion of any layer deposited on them. Indeed, the actual process of depositing the layer may alter their surface nature. A prominent factor influencing the strength of adhesion is the degree of chemical compatibility between the polymer chains at the surface of each fibre and the molecules comprising the deposited conducting layer. For example, polypropylene and polyethylene are quite inert towards other species because they are in essence polymeric hydrocarbons, so adhesion of a conductive layer to the surfaces of these fibres will be weak unless the surfaces are first suitably treated. By contrast, stronger adhesion would be expected to the surfaces of cellulosic fibres, such as viscose or washed and bleached cotton. The morphological nature of a fibre surface also has to be taken into account.

The uneven, scaly nature of the wool fibre surface would render the uniform deposition of thin layers particularly challenging, whereas synthetic fibres generally possess much smoother surfaces.

3.3 FABRIC CONSTRUCTIONS

Yarns are then converted to fabrics by means of weaving, knitting or embroidery, or by a non-woven process for the formation of webs and felts. As will be discussed later in this chapter, some fabric constructions are inherently more suited than others for adaptation as PV fabrics, depending in some measure on the method by which they are made PV (discussed in Chapter 5). The degrees of yarn bending and yarn interlacing in a fabric are particularly important in this respect. So too is fabric count, the number of warp and weft yarns per unit area of fabric. A description of each type of fabric construction now follows; fuller accounts are available in standard textbooks on textiles (Hatch, 1993; Sinclair, 2015).

3.3.1 WOVEN FABRICS

Most of us would recognise woven fabrics when we encounter them. They constitute textile structures in which there are usually two sets of yarns (though sometimes more) positioned at right angles to each other. Yarns that traverse in the lengthwise direction are commonly known as warp yarns and those in the crosswise direction as weft yarns. Where warp and weft yarns cross one another, they are very often interlaced. Interlacing yarns swap position from the top side to the underside of the fabric, and vice versa. There are three basic woven structures: plain, twill and satin, as illustrated in Figure 3.4. Figure 3.5 shows cross-sectional views of these woven structures. More complex woven structures have also been devised, but these are not discussed here.

At the points where the warp and weft yarns interlace, they bend considerably. Whilst the centre of the bent yarn is largely free of any distortion, the convex part is stretched and the concave part is compressed, as illustrated in Figure 3.6.

Woven fabrics are generally more rigid than other textile fabrics and possess greater dimensional stability. They are thus nearest to accommodating the rigidity and brittleness of PV and conducting layers deposited on them.

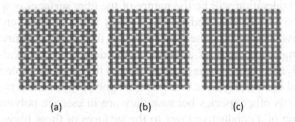

(a)   (b)   (c)

FIGURE 3.4 Basic woven structures. (a) Plain weave. (b) Twill. (c) Satin. (We are grateful to Dr Danmei Sun of Heriot-Watt University (Scottish Borders Campus, Galashiels, Scotland) for providing this schematic diagram.)

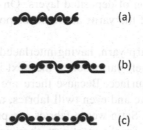

(a)

(b)

(c)

FIGURE 3.5 Cross-sectional views of woven structures. (a) Plain weave. (b) Twill. (c) Satin.

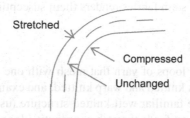

Stretched

Compressed

Unchanged

FIGURE 3.6 Schematic illustration of extension and compression in bent yarn.

Plain weaves constitute the most common type of weave structure. In a basic plain weave, the warp and weft yarns interlace at every crossing, and so the appearances of the top side and underside are identical. This high incidence of interlacing points has a substantial influence on the performance of plain weave structures when compared to those of twill and satin fabrics with identical yarns and fabric counts. Both sides of a plain weave fabric are relatively flat compared with those of other weave structures and are arguably the most suitable for printing and other surface finishes. This consideration will be important for the deposition of PV layers onto the fabric. It can also be noted that in plain weaves, compared with the other weave structures, interlacing at every crossover point works to reduce the fabric count, in that it precludes the very close proximity of adjacent parallel yarns. The result is that there will be greater fabric porosity than in other weaves constructed from the same yarns, and therefore more scope for incident light to pass right through the fabric.

In twill weaves, the yarns interlace after two or more crossings. As a result, there are a series of diagonal lines on at least one side of the fabric, which give rise to ridges on the fabric surface. These ridges will render the even deposition of thin PV and conducting layers more difficult to achieve, and it is perhaps worth noting here that twill fabrics are seldom printed. In many cases, the twill lines lie at 45°. Twill fabrics are noted for their durability and tend to feel softer than comparable plain weave fabrics. In addition, twill fabrics tend to possess higher strength than can be achieved in comparable plain weave structures, because the fewer interlacing points allow greater proximity of adjacent weft and warp yarns. The fabric count is higher. A problem with twill fabrics is their propensity to develop shine at points where abrasion is greatest. Abrasion tends to flatten the diagonal ridges, and more light is consequently reflected. Twill fabrics are also prone to twisting, which could lead to cracking and delamination of deposited layers. On the other hand, there is least bending and compression of the yarns at crossover points, a factor providing some mechanical stability.

In satin weaves, each warp yarn, having interlaced over and under a weft yarn, then crosses two or more weft yarns before the next interlacing. This construction produces a smooth fabric surface. Because there are so few interlacing points in comparison with plain-weave and even twill fabrics, adjacent parallel yarns can be packed very closely indeed. Satin weave fabrics are produced largely for their aesthetic quality as in practice, for the most part, they consist of very fine yarns. As a result, they possess appreciably lower porosity than equivalent plain-weave fabrics. Despite the durability that might be expected from the high-fabric count that can be achieved, the structure of satin fabrics renders them susceptible to abrasion.

3.3.2 KNITTED FABRICS

Knitted fabrics comprise loops of yarn that mesh with one another. There are two basic knit structures: weft knitted and warp knitted, and examples of both are shown in Figure 3.7. In the more familiar weft-knitted structure (used for example in hand knitting), a continuous yarn feeds through consecutive loops in a transverse direction. Warp-knitted structures are produced using multiple yarns, usually filament yarns, and the loops intermesh diagonally with adjacent columns. Both types of structure are highly porous: the structure provides a less stable base for a deposited layer, and much of the light that penetrates as far as the fabric is transmitted through it. Moreover, in both types of structure, the yarns are subjected to tight bends with small radii of curvature. The yarns may therefore be rendered more susceptible to rupture than they would be in a woven structure. Yarn breakage in a weft-knitted fabric results in laddering.

Despite the variety of knitted fabrics that can be constructed, a few overall observations can be made about their properties and hence their potential suitability as

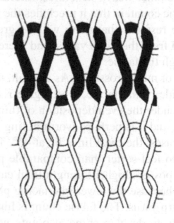 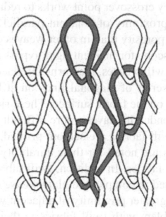

FIGURE 3.7 Knitted textile constructions: from left to right: plain weft knit stitch; warp knit tricot stitch.

substrates for PV devices. Weft-knitted fabrics possess good flexibility and drape and also good resilience. Resilience describes the ability of textile fibres to recover from compressive deformations. However, their potential suitability as substrates for PV devices is hampered by their low-dimensional stability which, as already noted, is important for continuity in the deposited layers. Whereas the extensibility of weft-knitted fabrics is clearly desirable for clothing, for example, it will tend to promote fragmentation and delamination of the rigid films that are deposited on them. Warp-knitted structures will arguably perform better as substrates, as they are more dimensionally stable. Indeed, a few types possess dimensional stability quite close to that of woven structures. However, they generally lack the degrees of flexibility and drape that weft-knitted structures possess.

One way of achieving greater dimensional stability in knitted fabrics is by means of 'laying in' extra yarns. An inlaid-knitted fabric consists of a knitted structure into which other yarns, not knitted, have been incorporated, or laid in (Ray, 2012). In a warp-knitted fabric, the inlaid yarns are quite straight: there are no knitted loops. In a weft-knitted fabric, the inlaid yarns are often more wavy, so that the yarns can be accommodated within the structure. Inlaid yarns can be added to introduce desired design effects, but they are present as well in many compression-knitted fabrics for healthcare (Liu et al., 2013). Conductive yarns can also be laid in knitted fabrics (El-Sherif, 2005), as will be shown in Chapter 4. There is scope too for PV yarns to be incorporated in a similar way.

Another way of achieving greater dimensional stability lies in the use of some types of double weft-knitted structure, which possess two inseparable layers of loops. (Care is necessary for the choice of construction, however, as some double weft-knitted structures are designed instead for their stretch properties!) Although many are considered too stiff for clothing, these features would render the double weft-knitted fabrics as more suitable substrates for PV devices. Nevertheless, they still possess quite open structures such that light can quite readily pass straight through them.

In general, then, most knitted structures appear less suitable than woven structures for PV devices.

3.3.3 EMBROIDERY

Interestingly, and perhaps unexpectedly, embroidered structures are increasingly being used as constructions for conductive textiles, and there may therefore be scope in the future for their adoption as a type of PV textile construction. This possibility will be discussed later on in the book, so it behoves us here to consider briefly the basis of embroidered structures. Embroidery has traditionally been perceived as craftwork. Indeed decorative silk embroidery has apparently been practised in China for thousands of years. At their most basic, embroidered structures comprise a collection of yarns applied to a base fabric or other material that achieves a desired pattern or design.

Embroidery is a textile technology in which each yarn or thread can be placed in almost any direction, and with the advent of advanced computer technology and computer-driven machinery, highly complex patterns can be produced.

As a consequence, embroidered structures have increasingly been devised for high-tech applications, and indeed structures can be specifically customised for an individual application. Examples in medical technology include surgical implants, such as stents for repairing aneurysms and supports for shoulder repairs. In mechanical engineering, patterns can be designed of high-performance filament yarns, like Kevlar® and carbon fibre, such that they are positioned along the lines of force acting on them. In electrically conducting textiles, embroidered structures can be fabricated that serve as sensors or antennae attached to a fabric. It may therefore also be possible – and useful – to fabricate an embroidered pattern of conductive threads on a textile fabric substrate for other applications, or even to construct a pattern of PV threads onto a fabric.

Three methods are used for embroidering sensors: chain stitch, standard embroidery and tailored fibre replacement (TFP). Only a brief overview is given here; more detailed accounts of each method can be found elsewhere (Bosowski et al., 2015). Chain stitch uses a single thread and has some similarities to crochet. For constructing sensors, moss embroidery is generally used which, as its name suggests, produces a pattern of moss-like structures. A disadvantage of the method is that, as there is only a single thread, a break in the thread is liable to produce a break in the entire system, similar to laddering in a weft-knitted textile. The method also produces a notably more textured fabric surface.

Standard embroidery is a two-yarn system and includes a double-lock stitch. An upper and lower thread are sewn either side of the base fabric on which the embroidered pattern is to be created (Figure 3.8). The lower thread secures the top embroidery thread to the fabric, and the fabric is held on a frame under tension. The frame is programmed to move in the x- and y-directions, in order to obtain the desired pattern. The tension applied to the fabric improves the accuracy of the embroidery and ensures neater stitches.

The TFP method, devised in the 1990s, extends standard embroidery by using a three-yarn system. This method allows for the positioning of threads, such as sensor threads, in a particularly versatile and precise manner to yield a desired pattern using a highly controlled geometry. The sensor threads are secured by upper and lower threads onto a base fabric (Figure 3.9). The TFP technique has successfully been applied to the positioning of composite materials to fit predetermined loading conditions.

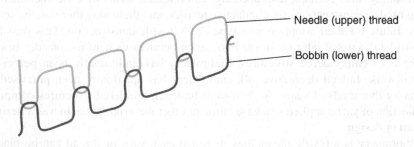

Needle (upper) thread

Bobbin (lower) thread

FIGURE 3.8 Standard embroidery – double-lock stitch. (From Dilak, T., *Electronic Textiles: Smart Fabrics and Wearable Technology* (Woodhead Publishing, 2015).)

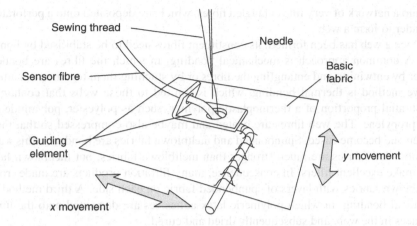

FIGURE 3.9 The basis of tailored fibre replacement (TFP) technologies. (From Dilak, T., *Electronic Textiles: Smart Fabrics and Wearable Technology* (Woodhead Publishing, 2015).)

3.3.4 NON-WOVEN FABRICS

There is also another class of fabric structures that we should include. They are quite distinct from those already discussed and also quite enigmatic in that they are difficult to define exactly. These structures are collectively known as non-woven fabrics, although opinions still differ over whether certain structures are non-woven or not. The term, non-woven, perhaps a little unhelpfully, tells us something about what these fabrics are not rather than about what they are. In addition to not being woven fabrics, they are also neither knitted nor embroidered fabrics.

So what are they? Although formulating anything like a formal definition of non-woven fabrics is challenging, an indicative description is that they are structures formed from short textile fibres, continuous filaments or staple, which have been constructed into webs and then bonded together. Thus, fibrous webs are the basis of non-woven textiles. There are many different types of non-woven fabric, and a detailed discussion of them all would be beyond the scope of this book. Good authoritative accounts of the structures and properties of non-woven fabrics are available elsewhere (Russell, 2007). Nevertheless, despite their variety, non-woven webs can usefully be categorised as drylaid, wetlaid or polymer-laid (also known as spunmelt).

Drylaid non-wovens are produced from staple fibres. The fibres may first be carded (combed) to produce a thin web, which is deposited onto a continuously moving conveyor. Alternatively, they may be airlaid, whereby fibres are suspended in air and then transported to an air permeable conveyor.

Wetlaid webs are produced by processes similar to those used for producing paper. The fibres are first dispersed in water and then collected on a continuous screen to form a uniform mesh. The water is then filtered off before the web is dried by heating.

Polymer-laid webs can be formed from continuous filaments, which are collected on a continuous conveyor belt. This is the basis of the production of spunbonded non-woven fabrics. In the meltblown process, freshly extruded molten polymer is broken up into a network of very fine entangled fibres, which are deposited onto a perforated cylinder to form a web.

Once a web has been formed, its constituent fibres need to be stabilised by bonding. A common approach is mechanical bonding, in which the fibres are bonded either by entwining and entangling the fibres or by stitching them together. An alternative method is thermal bonding, which is limited to those webs that contain a substantial proportion of a thermoplastic polymer, such as polyester, polyamide or polypropylene. The web fibres are heated and maybe also compressed so that they soften and become fused. Spunbonded and meltblown fabrics are bonded in this way. Spunbonded fabrics are much stronger than meltblown fabrics, but meltblown fabrics make excellent filters. In consequence, many filtration products are made from meltblown fabrics with layers of spunbonded fabric on each side. A third method is chemical bonding, in which polymeric latex adhesives are deposited onto the fibre surfaces in the web, and subsequently dried and cured.

Finally, non-woven fabrics may undergo a finishing process, influenced by the application for which they are destined. For example, wet finishing may be through impregnation with fragrances, perfumes or medical formulations for cosmetics and biomedical industries. Other finishes confer antistatic properties and water repellency. Dry finishing techniques include embossing and micro-corrugation. Such finishes, particularly those affecting the surface of a non-woven fabric, will however almost certainly influence the effectiveness of a non-woven fabric as a PV substrate.

It therefore becomes apparent that non-woven fabrics possess a huge range of properties, depending *inter alia* on the properties of the constituent fibres within a web and the type of bonding holding the fibres together. Moreover, it is difficult to achieve high strength along with high flexibility. One property, however, that all non-woven fabrics possess is high porosity.

3.4 SOME CONCLUDING REMARKS

We can see that textile fabrics do indeed encompass a myriad of constructions. Even within each of the four main types of construction that have been discussed in this chapter, very many different fabrications are possible. In addition, there are choices in the physical and chemical natures of the fibres that can be exploited in the construction of these fabrics. It is not surprising therefore that in devising any solar textile, we need to take account of the properties of the fabric that will support the PV cells.

REFERENCES

Bosowski, P., Hoerr, M., Mecnika, V., Gries, T. & Jockenhövel, S. (2015). Design and manufacture of textile-based sensors. In *Electronic Textiles: Smart Fabrics and Wearable Technology*, Dias, T. (Ed.), Woodhead Publishing, Cambridge, UK, pp 75–107.
El-Sherif, M. (2005). Integration of fibre optic sensors and sensing networks into textile structures in wearable electronics and photonics. In *Wearable Electronics and Photonics*, Tao, X. (Ed.), Woodhead Publishing, Cambridge, UK, pp. 105–135.
Hatch, K.L. (1993). *Textile Science*. West Publishing Company, Minneapolis, MN.

Jones, F.R. (2001). Glass fibres. In *High-performance Fibres*, Hearle, J.W.S. (Ed.), Woodhead Publishing , Cambridge, UK, pp. 191–238.

Liu, R., Lao, T.T. & Wang, S.X. (2013). Impact of weft laid-in structural knitting design on fabric tension behavior and interfacial pressure performance of circular knits, *Journal of Engineered Fibers and Fabrics*, 8, 96–107.

Mather, R.R. & Wardman, R.H. (2015). *The Chemistry of Textile Fibres*, 2nd ed., Chapter 6, Royal Society of Chemistry, London, UK.

Ray, S.C. (2012). *Fundamentals and Advances in Knitting Technology, Chapter 6*. Woodhead Publishing India Pvt Limited, Delhi, India.

Russell, S.J. (Ed.) (2007). *Handbook of Nonwovens*, Woodhead Publishing, Cambridge, UK.

Sinclair, R. (Ed.) (2015). *Textiles and Fashion: Materials, Design and Function*, Woodhead Publishing, Cambridge, UK.

4 Strategies for Achieving Electrically Conducting Textile Fabrics

4.1 INTRODUCTION

It can be observed from Figure 2.1 in Chapter 2 that the top and bottom of a solar cell need to be electrically conducting. In particular, the surface of the supporting substrate needs to be rendered conducting before semiconducting layers are deposited on it to provide a PV device. As we have seen, textile fabrics are unlike other PV substrates in not being homogeneous, with the consequence that there are several pathways by which PV properties can be conferred. Most of these pathways take account of the requirement that conductivity must initially be conferred on the fabric.

One strategy is first to make the constituent yarns conducting. The conducting yarns are constructed into fabric, and solar cells can then be deposited on them. This method involves the production of specially customised fabrics. Alternatively, conductivity is conferred on the already constructed fabric. This alternative approach can be achieved either by interlacing conducting yarns into the fabric or by first depositing a thin continuous conductive coating onto the fabric before solar cells are deposited onto it. The latter method has the advantage that it can be applied to 'off-the-shelf' fabrics. Whichever method is adopted, the surfaces of the yarns constituting the fabric must be rendered conductive, to enable the performance of the solar cells that will be deposited on them. Still another approach to achieving a solar textile fabric is to render the individual yarns electrically conducting and then make them PV *before* fabric construction. These pathways are summarised in Figure 4.1. The relative merits of these strategies are discussed at length in Chapter 5. The present chapter aims to consider how electrical conductivity can first be obtained in an actual textile fabric before it is rendered PV.

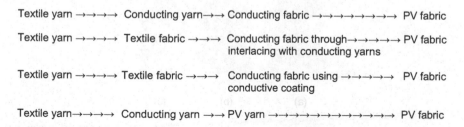

FIGURE 4.1   Pathways to the construction of PV fabrics.

DOI: 10.1201/9781003147152-4

4.2   CONDUCTING YARNS

At first sight, the simplest approach to achieving electrical conductivity in yarns is to use yarns that are already inherently conductive. An obvious candidate would be a very thin metal yarn, and indeed thin metal yarns are produced with diameters less than 40 µm (Stoppa & Chiolerio, 2014). Whether a metallic fabric constructed directly from such yarns is classed as a textile is probably more of a subject for esoteric discussion, although metallic fabrics are manufactured commercially, for example for the fashion sector.

A related approach is to combine metal yarns with other yarns during fabric construction. Traditional yarn spinning techniques, such as ring-spinning, can be adapted to achieve these yarn blends. The yarns must be thin enough not to compromise the flexibility of the fabric significantly, but on the other hand thin metal yarns are more liable to break during fabric construction and application. Moreover, the risk exists that the necessary contacts between the metal yarns will in fact only be intermittent.

Three types of blended yarn can be broadly identified: those with metal yarn twisted around the conventional yarn, those with the conventional yarn twisted around the metal yarn and braided structures in which the yarns are twisted around one another. These different types are shown schematically in Figure 4.2; here the yarns are packed together as closely as possible. The first of these types of blend, shown in Figure 4.2b, could be attractive in that, provided the outer metallic sheath is not degraded or abraded during subsequent fabric construction, such a blend would ensure a good electrically conducting yarn surface on which to deposit PV layers. The risk of only intermittent connectivity between yarns, however, remains. Blended yarns of this type are hardly new. Winding very thin strips of precious metals around silk or linen yarns to produce decorative yarns was practised at least 2000 years ago. Fabrics from such yarns were considered luxury items. Indeed, reference can be found in the Bible to the technique of decorating vestments with precious metals (Exodus, Chapter 39, Verses 2–3), in ancient India saris could be adorned with gold ornamentation and examples from several other parts of the ancient world are also known.

The second type of blend, shown in Figure 4.2a, would be less effective, although generally easier to engineer. The metallic core is largely hidden, and hence the

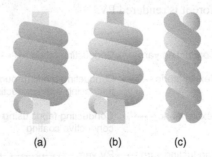

(a)   (b)   (c)

FIGURE 4.2   Some structures of blended yarns, consisting of conventional and metal yarns (a–c). (From Dilak, T., *Electronic Textiles: Smart Fabrics and Wearable Technology* (Woodhead Publishing, 2015).)

surface of the blended yarn is not conducting. In addition, the blended yarn is less flexible and scarcely extensible. It is therefore difficult to weave or knit in standard looms and knitting machines. Blended yarns based on braided structures, as shown in Figure 4.2c, can be produced simply by twisting conventional and metal yarns together, but these also generally possess poor extensibility (Qu & Skorobogatiy, 2015).

Yarns can alternatively be made from polymers that possess electrical conductivity. Examples are polypyrrole, polyaniline and polythiophene. Nowadays, the conductive polymer of choice is very often PEDOT:PSS, which is a salt made up of positively charged poly(ethylenedioxythiophene) (PEDOT) and negatively charged polystyrene sulfonate (PSS), whose structure is shown in Figure 4.3. However, the conductivities of these polymers are 2–3 orders of magnitude lower than those of many metals, they lack the mechanical strength of conventional yarns and their flexibility is quite limited. Some yarns consisting of blends of conducting polymers with conventional polymers have been quite successful in cases where the conducting polymer has been able to withstand the processing conditions for producing yarn and subsequently for fabric construction (Akbarov et al., 2005).

Instead of applying blending processes, metal or conductive polymer can be deposited onto a yarn surface. The yarn surfaces then acquire good conductivity, yet the bulk properties of the yarn are unchanged. Metals can be coated onto yarns by a number of techniques, including vacuum evaporation and sputter coating. In the former technique, a source of the metal is heated *in vacuo*. Because the procedure is carried out *in vacuo*, metal atoms that have evaporated from the source are deposited directly on the yarns, without hindrance from atoms in the surrounding medium. Eventually, a thin metal coating is formed on each yarn. In the sputtering technique, a metal target is bombarded by energetic ions. As a result, metal atoms are ejected from the target, and if they have acquired sufficient kinetic energy, they can travel

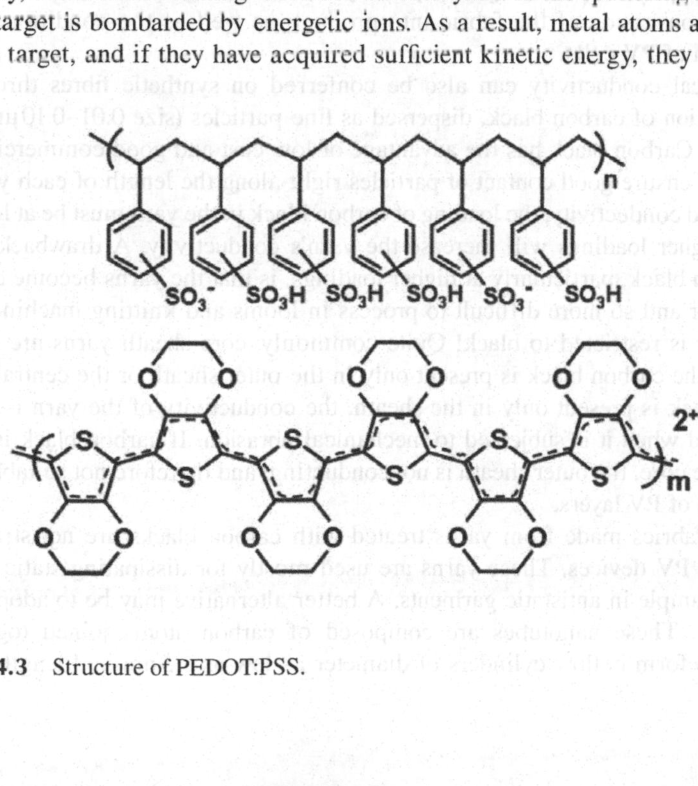

FIGURE 4.3   Structure of PEDOT:PSS.

to the yarns. There they form a metallic film. Good accounts of both techniques are available in a book chapter by Wei et al. (2009).

The deposition of a conductive polymer may be achieved by a number of means. The conductive polymer may, for example, be deposited from suspension. Alternatively, the conductive polymer can in some cases be formed by polymerisation in solution or suspension in the presence of the textile yarns (Malinauskas, 2001). As polymerisation proceeds, the resulting polymer is precipitated onto the yarns' surfaces. To achieve deposition of the conducting polymer, the precursor monomer can be polymerised in solution or in a suspension in the presence of the yarn. Better control can often be achieved though where the monomer is polymerised on the actual yarn surfaces. After the monomer has been adsorbed on the surfaces, it is polymerised through exposure to a suitable polymerisation initiator, almost invariably an oxidising agent (Malinauskas, 2001). The success of this approach is governed by the amount of monomer adsorbed, how uniformly it has been adsorbed and the chemical resistance of the yarns to the initiator.

The achievement of all these deposition techniques also depends on how tenaciously the deposited metal or polymer adheres to the yarns. A metal layer will be less extensible than the yarn to which it is attached, so is prone to fracture or peeling when the yarn is flexed. The metal layer is no longer continuous and conductivity is lost. Moreover, evaporated metal films tend to adhere less strongly than sputtered films. Some resistance to fracture and peeling, however, appears to be conferred if the textile yarn is first coated with a layer of conductive polymer before deposition of the metal coating (Qu & Skorobogatiy, 2015). In addition, the layers of coating, whether metallic, polymeric or both, must withstand all the subsequent processes undergone by the yarn, including fabric construction and then the conversion of the fabric into product, as well as the conditions for the deposition of PV cells.

Electrical conductivity can also be conferred on synthetic fibres through the incorporation of carbon black, dispersed as fine particles (size 0.01–0.10 μm) along the yarns. Carbon black has the advantage of low cost and good commercial availability. To ensure good contact of particles right along the length of each yarn, and hence good conductivity, the loading of carbon black in the yarn must be at least 10% (w/w). Higher loadings will increase the yarn's conductivity. A drawback of adding carbon black, particularly at higher loadings, is that the yarns become considerably stiffer and so more difficult to process in looms and knitting machines. Also, the colour is restricted to black! Quite commonly, core-sheath yarns are adopted, in which the carbon black is present only in the outer sheath or the central core. If carbon black is present only in the sheath, the conductivity of the yarn is liable to be reduced when it is subjected to mechanical abrasion. If carbon black is present only in the core, the outer sheath is not conducting, and therefore not suitable for the deposition of PV layers.

Thus, fabrics made from yarns treated with carbon blacks are not strong candidates as PV devices. These yarns are used mostly for dissipating static electricity, for example in antistatic garments. A better alternative may be to adopt carbon nanotubes. These nanotubes are composed of carbon atoms joined together in clusters to form hollow cylinders of diameter as low as 1.5 nm and length >1 μm.

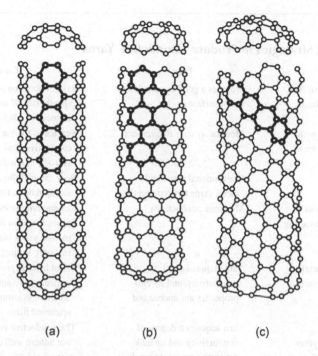

(a)   (b)   (c)

FIGURE 4.4 Structures of single-walled carbon nanotubes. (a) Armchair. (b) Zig-zag. (c) Chiral. (From Mather, R.R. and Wardman, R.H., *The Chemistry of Textile Fibres: Edition 2*; (The Royal Society of Chemistry, 2015).)

In single-walled nanotubes (SWNTs), the walls are formed from sheets of one molecule thickness, so-called graphene. There are rounded caps at both ends of each cylinder, as illustrated schematically in Figure 4.4. Multiwalled nanotubes also exist, which are essentially a collection of concentric SWNTs of different diameters. Although carbon nanotubes are more expensive than carbon blacks, the loading required to confer conductivity is generally much lower, 1%–5%. Carbon nanotubes also provide some reinforcement to yarns. More details about these fascinating materials are available elsewhere (e.g. Mather & Wardman, 2015).

The pros and cons of these techniques are summarised in Table 4.1.

4.3 CONDUCTING FABRICS

As stated in the Introduction, there are two broad approaches to conferring electrical conductivity on textile fabrics. One is the integration of conducting yarns in the fabric structure, and the other is the addition of a continuous conducting layer over the whole of one surface of the fabric. The integration of conducting yarns is important, for example, in connecting tiny sensors embedded in textile fabric structures. It is arguably the more complex process, and it leads to a less uniform conducting fabric surface. Moreover, the integrated yarns have to be interlaced with the yarns constituting the fabric and, therefore, like these yarns, are subjected to bending at

TABLE 4.1
Summary of Strategies to Produce Conductive Yarns

Method	Pros	Cons
Conventional yarn core with metallic sheath	Provides a good conducting yarn surface	Risk of abrasion of metallic sheath; risk of only intermittent connectivity between yarns
Metal yarn core with conventional sheath	Moderately easy to engineer	Surface of the yarn blend is not conductive; less flexible and scarcely extensible
Braids	Conventional yarns and metal yarns twisted together	Poor extensibility, so difficult to manipulate in later processing
Yarns formed from electrically conducting polymers	Moderate conductivity	Conductivity is still 100–1000 times less than for many metals; yarns lack mechanical strength; limited flexibility
Deposition of metals onto a yarn surface	Yarn acquires good conductivity, and its bulk properties are unchanged	Metal layer is prone to fracture or peeling. Evaporated films adhere less strongly than sputtered films
Deposition of conductive polymer onto a yarn surface	Yarn acquires a degree of conductivity and its bulk properties are unchanged	The conductive polymer may not adhere well enough to the yarn
Incorporation of carbon black	Low cost. Carbon black has good commercial availability	Loading must be at least 10% and so the yarns become considerably stiffer. Colour is restricted to black
Incorporation of carbon nanotubes	Good conductivity at much lower loadings (1%–5%) than with carbon black	More expensive than carbon black. Colour is restricted to black

interlacing points. Thus, those structures in which there is least yarn bending are likely to be the ones that are most desirable for incorporating conducting yarns.

In the case of woven fabric structures, some of the textile yarns in the warp and weft directions can be replaced by conducting yarns during construction of the fabric, as shown in Figure 4.5. It can be noted too that of the woven structures shown in Figure 3.5, yarns in the twill weave structure are subjected to less severe bending than those in the plain and satin weave structures. The best integration of conducting yarns is therefore in some types of twill weave structures.

Knitted fabrics present more of a problem, in that the yarns in both warp- and weft-knitted structures are bent much more severely at each loop. The bends are so tight that, depending on the nature of the conducting yarn, conductivity could be lowered due to partial fracture of the yarn. Complete fracture of the yarn would lead to total loss of conductivity. It is risky then to interlace the conducting yarns in the same manner as the component fabric yarns, especially if the conducting yarns include a metallic component. Instead the conducting yarns can be integrated into much straighter lines that interlace with loops of the knitted fabric (Figure 4.6).

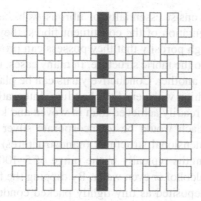

FIGURE 4.5 Conductive wires integrated into a woven fabric in the warp and weft directions. (From Tao, X., *Wearable Electronics and Photonics* (Woodhead Publishing, 2005).)

Optical fibre
or electrical wire

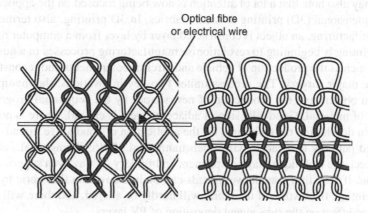

FIGURE 4.6 Conductive wire integrated into warp (left) and weft (right) knitted structures. (From Tao, X., *Wearable Electronics and Photonics* (Woodhead Publishing, 2005).)

The drawback to this approach is that the physical and aesthetic character of the fabric may be compromised to quite a large extent. The integrated yarns also have to be connected to one another in a continuous fashion, in order to ensure conductivity throughout the fabric. The processing of such structures can be quite challenging for knitting machines. Nevertheless, electrically conducting knitted fabrics are available on the market. These fabrics are produced by knitting machines that can integrate into these fabrics metal yarns or metal-coated synthetic yarns.

Embroidery may be exploited as a means of producing conducting fabrics because a desired arrangement of conductive yarns can be pre-programmed (Post et al., 2000). However, whilst this approach may be valuable for imparting conductivity, the fabric does become more textured, a feature that could make the uniform deposition of PV layers more challenging. To our knowledge, little has been attempted so far on rendering embroidered structures PV, but the approach may well turn out to have exciting possibilities.

Non-woven fabrics consist essentially of fibrous webs, as noted in Chapter 3. There is no overall orientation of the constituent fibres; they are more or less randomly arranged. The high porosity that they possess can provide scope for integrating conductive yarns into the fabric structure. In many cases, there are layers of short fibres laid on top of one another. Conductive yarns can be placed between these layers in a nearly linear fashion with very little bending. Integration of conductive yarns can then be much more readily achieved than in woven and knitted fabrics.

An alternative approach would be to incorporate silver nanowires. These are essentially tiny one-dimensional structures (nanostructures) with cross-sections of only 10–200 nm and lengths typically in the range of 5–100 μm. Silver nanowires can be dispersed in an alcohol or even water. By dipping the fabric into a dispersion, the nanowires can be deposited as tiny tightly packed conducting cylinders on the surfaces of the fibres comprising the fabric, and indeed within them. The conducting metal added to the fabric then consists of a whole host of tiny individual wires that touch one another.

We may also note that a lot of attention is now being focused on the application of three-dimensional (3D) printing to textile fabrics. In 3D printing, also termed additive manufacturing, an object is constructed layer by layer from a computer model of it. 3D printing is beginning to revolutionise manufacturing processes in a number of sectors, such as the production of vehicle and aircraft components and reconstruction in severe major surgery. The process unites two key innovations: the manipulation of digital objects and the production of new shapes by the sequential programmed addition of material. Good adhesion of adjacent layers is crucial. There is now keen interest in developing the technique for the application of conductive threads in programmed patterns to a fabric (Grimmelsmann et al., 2016; Zhang et al., 2019). In this respect, the technique may complement embroidery or may even supersede it. In particular, if very thin conductive threads can be laid down on the fabric by means of 3D printing, the texture of the fabric will hardly be altered, and there will then be little or no effect on the subsequent deposition of PV layers.

A very recent development in the construction of conductive textiles is the incorporation of narrow flexible electronic filaments of width less than 5 mm. Complex circuits can be devised on flexible plastic filaments, which are then incorporated into a textile fabric by conventional weaving or knitting processes (Komolafe et al., 2021). Although, to our knowledge, no examples have been produced on a commercial scale, a number of prototypes have been demonstrated at the research level.

We can now turn to the other approach to rendering textile fabrics electrically conducting: the deposition of a continuous conducting layer on one side of the fabric. The choice of a suitable material for the conducting layer will depend not only on its inherent conductivity and its uniform deposition on the fabric but also on its compatibility with the fabric and its compatibility with the PV layers that will be deposited on it. Moreover, a conductive layer must be able to survive flexing and twisting of the fabric without fracture, just as in the case with individual yarns. Again, metal layers or those comprising conducting polymers are the layers of choice.

A great many techniques for conferring conductivity on textile fabrics have been devised, and a few illustrative examples are given here. More detailed information can be obtained from a review by Tseghai et al. (2020). One approach to conferring conductivity is printing a conductive ink or paste onto the fabric. For example, conductive inks can be applied by screen printing, a historic process for the decorative printing of textile fabrics. Traditional screen printing involves pressing ink through a stencilled mesh screen to create a desired design. The screen was traditionally made of silk, though nowadays is usually woven polyester. To create a conducting fabric, a metal paste – commonly silver paste – is applied. The method allows only selected areas of the fabric to be made conductive if desired. The metal deposited on the fabric must be sufficiently thick to impart high enough conductivity, so several passes may be required. However, care has to be taken that after drying the deposited metal does not crack when the fabric is bent or folded. Indeed, screen-printed conductive textile fabrics are quite susceptible to stresses caused by bending and abrasion.

Inkjet printing can also be used effectively for printing conductive ink onto textile fabric. It is essentially a non-impact printing process, whereby small drops of ink are directed in rapid succession through tiny nozzles onto a substrate. Inkjet printers are familiar in the home and office, where the substrate of course is paper! Many conductive inks consist of minute particles of metal – size of the order of $0.01–0.10\,\mu m$ – dispersed in a suitable liquid, normally an aqueous one. The particles have to be minute so that they can pass freely through the nozzles. The particles also have to be finely dispersed but tiny, finely dispersed particles often have a tendency to aggregate. Aggregation of the metal particles would give rise to uneven deposition of metal on the fabric, as well as risking clogging in the inkjet nozzles. To impede aggregation requires the addition of auxiliary compounds, hence making the formulation of the ink more complex. Also, the viscosity and surface tension of the ink must both be sufficiently low to allow free passage through the nozzles.

The ink is deposited on the fabric as discrete droplets, which on drying give rise to minute adjacent zones of metal particles. To provide continuous connectivity between them on the fabric surface, these particles may then have to be sintered. Sintering is a process in which a mass of solid material is compacted by heat or pressure below its melting point. However, with the elevated temperatures normally required come the risks of fabric degradation and fracture of the deposited metal layer. For example, tiny silver particles can be sintered at 180°C (Stoppa & Chiolerio, 2014). Commodity fabrics made from polyester would therefore be unaffected as polyester melts at 250°C–260°C. On the other hand, fabrics such as those made from polypropylene, which melts at 160°C–165°C, would be degraded.

Clogging of the nozzles in the printer can be overcome if the ink is a true solution rather than a dispersion of tiny metal particles. To this end, there is now great interest in developing so-called particle-free inks (Bei et al., 2018; Yang et al., 2019). With these inks, a solubilised metal salt is chemically reduced to produce a smooth conductive layer of the metal on the fabric surface. The layer can be so thin that it is transparent, yet its conductivity can be as high as 80% of that of the bulk metal. Silver salts are commonly used, although in view of the high price of silver other metal salts, such as copper salts, are also being tested. In addition to the metal salts, the inks contain a chemical reducing agent, such as an organic acid, to reduce the salt to the metal, and often a complexing agent as well, such as ammonia or an organic amine, to improve the solubility of the metal salt. Reduction of the metal salt is triggered by heat, light or microwaves. In the case of silver acetate, the deposited metal film can be reduced at temperatures as low as 90°C. Thus, not only are these inks free of complications from particle aggregation, but they have the potential to be applied to standard commodity fabrics, including those made from a number of natural fibres. No doubt, improved formulations will be devised in the next few years on a commercial basis. Meanwhile, there has recently been an interesting development in which it has been shown that inkjet printing can confer a coating of silver on individual yarns in woven and even knitted polyester fabrics, such that yarn mobility is still maintained, yet a conductive network is created across the textile fabric structure (Shahariar et al., 2019). The aesthetic qualities and mechanical properties of the fabrics appear largely unaffected.

An alternative to inkjet printing is spray coating. This technique is, arguably, more versatile than inkjet printing in that it is much less sensitive to the viscosity of the ink that is being applied. The basic principle of spray coating is to use a spray gun to form minute droplets of an ink dispersion or solution to enable the formation of thin metallic films on a substrate. In the case of textile substrates, an interface layer of resin may first be applied to a fabric by screen printing, with the aim of filling in pores in the fabric and so presenting a smoother layer to support a spray-coated metal film. This combined approach will however reduce the quality of the fabric handle, though this will be less problematic for some applications, such as tents and canopies, than for others, such as clothing.

Another option is coating with a doctor blade (Figure 4.7). A sharp blade is placed at a fixed distance from the surface of the substrate to be coated (normally 10–500 μm). The ink is then placed in front of the blade. As the blade is moved across the substrate, a thin film of ink is deposited uniformly on the substrate. The thickness of the film is governed by a variety of factors: the distance between the blade and the substrate, the surface energy of the substrate and the surface tension and viscosity of the ink. Although optimisation of the conditions for applying an ink can be quite laborious, the technique can be readily adopted on an industrial scale. A related technique is slot-die coating, whereby an ink is fed to a substrate through a thin slit. Proponents of slot-die coating maintain that the technique gives better reproducibility of coating.

Still another option, which we are adopting in our work, is the deposition of a thin metallic film directly on the fabric by sputtering or vacuum evaporation

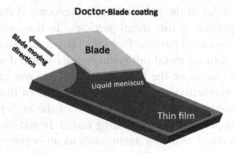

FIGURE 4.7 Schematic diagram showing doctor-blade coating (Courtesy of Aveena Abee Varghese and Lethy Jagadamma, St Andrews University).

(Diyaf et al., 2014). The techniques that can be used are the same as those already highlighted earlier in this chapter for deposition on yarns. Whilst this approach can certainly produce electrical conductivity at the fabric surface, care has nevertheless to be taken that the whole surface is covered. This consideration is especially important for a textured surface like that of a textile fabric. If some parts of the surface are not to be rendered conducting, they can be masked before the metal film is deposited. Moreover, as already highlighted in this chapter, the deposited metal film is liable to fracture when the fabric is bent or folded, with consequent loss of conductivity. To mitigate this problem, we first deposit a thin layer of conducting polymer such as PEDOT:PSS and then deposit a thin aluminium layer by vacuum evaporation. The partial resistance of these layers to fracture and peeling conferred by this strategy has been noted above (Qu & Skorobogatiy, 2015). Since the polymer is more flexible and more compatible with the fabric surface, there will still be enough electrical conductivity across any cracks that may subsequently appear in the aluminium film (Diyaf et al., 2014). As mentioned above, the conducting layers must be able to withstand all the processing and use to which the fabric will then be subjected.

A process that has great potential for the textile sector is electroless plating, a technique that is already used for example in barrier curtains for the construction industry and in fabrics for shielding against electromagnetic and radio frequency interference. Textile designers are also now recognising aesthetic applications of the process. Unlike electroplating, the process requires neither conducting electrodes nor an externally generated electrical supply. Instead, the process depends on the presence of a chemical reducing agent, normally sodium hypophosphite (NaH_2PO_2) which, in the presence of a suitable catalyst, reduces metal ions to metal atoms that are then deposited on the fabric surface. The metal most commonly used in electroless plating is nickel, but copper has also been used. A significant advantage of the process is the creation of an even layer of metal across the fabric surface irrespective of the fabric's texture. A detailed discussion of the electroless plating of textiles has been written by Jiang and Guo (2009).

Nevertheless, the continuous metal coating that has been deposited is still liable to fracture. The incorporation of metal nanowires may serve to overcome this problem. When the fabric is bent or stretched, fracture of the incorporated metal is arguably less likely to occur, as the individual nanowires would behave independently of one another to a large extent. Thus, electrical conductivity is more likely to be maintained.

The pros and cons of these techniques are summarised in Table 4.2.

4.4   SOME CONCLUDING REMARKS

It is apparent that, because of the inhomogeneous nature of textile fabrics, there are many ways in which electrical conductivity can be conferred on them. In particular, there is scope for making yarns conductive before the actual fabric is constructed. Delaying conferment of conductivity until the fabric has been constructed, however, does enable the adoption of 'off-the-shelf' fabrics. Whichever approach is chosen, several techniques are available, involving the addition of metal or conducting polymer or both, and the potential of some of these techniques has probably not yet been

TABLE 4.2

Summary of Strategies to Produce Conducting Fabrics

Method	Pros	Cons
Integration of conducting yarns in the fabric structure	Important in connecting tiny sensors embedded in a fabric. Potential to explore embroidery and 3D printing techniques and the application of silver nanowires	Gives a less uniform conducting fabric surface than deposition of a continuous conducting layer. The integrated yarns are subject to bending at interlacing points
Addition of a continuous conducting layer over the fabric surface:		
Screen printing	Well-established technique for textile fabrics	Deposited metal layers may crack when fabric is bent or folded
Inkjet printing	Now an established technique for textile fabrics	Aggregation of metal particles in inkjet nozzles – interest in circumventing problem using conductive inks that are true solutions
Spray coating	Much less sensitive to ink viscosity than inkjet printing	An interface layer of resin should first be applied to the fabric
Doctor blading	Readily adopted at an industrial level	Optimisation of application conditions can be laborious
Direct deposition of a thin metallic layer on the fabric surface	Established techniques such as sputtering or vacuum coating can be used	Textured nature of the fabric surface can be problematic. The film may break or peel when the fabric is bent or folded – less problematic if a layer of conductive polymer has first been deposited
Electroless plating	Creates an even metallic layer irrespective of fabric texture	The deposited metallic layer may fracture

fully realised. The special nature of textile fabrics allows immense versatility in the strategies for achieving electrical conduction.

REFERENCES

Akbarov, D., Baymuratov, B., Akbarov, R., Westbroek, F., de Clerck, K. & Kiekens, P. (2005). Optimising process parameters in polyacrylonitrile production for metallization with nickel, *Text. Res. J.*, 75, 197–202.

Bei, Y., Cheng, H. & Zu, M. (2018). Research status and prospects of particle-free silver conductive ink, *IOP Conf. Series: Materials Science and Engineering*, 394, 042060 (8 pages).

Diyaf, A.G., Mather, R.R. & Wilson, J.I.B. (2014). Contacts on polyester textile as a flexible substrate for solar cells, *IET Renew. Power Gener.*, 8, 444–450.

Grimmelsmann, N., Martens, Y., Schäl, P., Meissner, H. & Ehrmann, A. (2016). Mechanical and electrical contacting of electronic components on textiles by 3D printing, *Procedia Technology*, 26, 66–71.

Jiang, S.Q. & Guo, R.H. (2009). Modification of textile surfaces using electroless deposition. In *Surface Modification of Textiles*, Wei, Q. (Ed.), Woodhead Publishing, Cambridge, UK, pp. 108–125.

Komolafe, A., Zaghari, B., Torah, R., Weddell, A.S., Khanbareh, H., Tsikriteas, Z.M., Vousden, M., Wagih, M., Jurado, U.T., Shi, J., Yong, S., Arumugam, S., Li, Y., Yang, K., Savelli, G., White, N.M. & Beeby, S. (2021). E-textile technology review – From materials to application, *IEEE Access*, 9, 97152–97169.

Malinauskas, A. (2001). Chemical deposition of conducting polymers, *Polymer*, 42, 3957–3972.

Mather, R.R. & Wardman, R.H. (2015). *The Chemistry of Textile Fibres*, 2nd Edition, Royal Society of Chemistry, Cambridge, UK, pp. 252–254.

Post, E.R., Orth, M., Russo, P.R. & Gershenfeld, N. (2000). E-broidery: design and fabrication of textile-based computing, *IBM Systems J.*, 39, 840–860.

Qu, H. & Skorobogatuy, M. (2015). Conductive polymer yarns for electronic textiles. In *Electronic Textiles: Smart Fabrics and Wearable Technology*, Dias, T. (Ed.), Woodhead Publishing, Cambridge, UK, pp. 21–53.

Shahariar, H, Kim, I., Soewardiman, H. & Jur, J.S. (2019). Inkjet printing of reactive silver on textiles, *ACS Appl. Mater. Interfaces*, 11, 6208–6216.

Stoppa, M. & Chiolerio, A. (2014). Wearable electronics and smart textiles: A critical review, *Sensors*, 14, 11957–11992.

Tseghai, G.B., Malengier, B., Fante, K.A., Nigusse, A.B. & van Langenhove, L. (2020). Integration of conductive materials with textile structures, an overview, *Sensors*, 20, 6910 (28 pages).

Wei, Q., Xu, Y. & Wang, Y. (2009). Textile surface functionalization by Physical Vapor Deposition (PVD). In *Surface Modification of Textiles*, Wei, Q. (Ed.), Woodhead Publishing, Cambridge, UK, pp. 58–90.

Yang, W., List-Kratochvil, E.J.W. & Wang, C. (2019). Metal particle-free inks for printed flexible electronics, *J. Mater. Chem.*, 7, 15098–15117.

Zhang, M., Zhao, M., Jian, M., Wang, C., Yu, A., Yin, Z., Liang, X., Wang, H., Xia, K., Liang, X., Zhai, J. & Zhang, Y. (2019). Printable smart pattern for multifunctional energy-management E-textile, *Matter*, 1, 168–179.

5 Enabling Textile Fabrics to Become Photovoltaically Active

5.1 INTRODUCTION

At this point, it will be appreciated that making a piece of fabric into an active converter of solar illumination into electricity has some obvious difficulties. Instead of the smooth planar surface of a sheet of glass, or even of a plastic film, on which to attach the various layers of a photovoltaic device, we have the uneven textured, porous, and sometimes soft, surface of, say, a woven or felted material. Whilst the previous chapter has shown that there are manufacturing techniques for overcoming some of these complications and also for ensuring that we can start with an electrically conducting surface, the challenge of transferring standard semiconductor fabrication methods to a deformable base having temperature constraints is immense. Even when this has been achieved, the performance will be constrained if the coatings are not sufficiently well bonded to resist cracking or delamination when flexed. The options for producers all have pros and cons with no outright winner to please all users.

5.2 ALTERNATIVE STRATEGIES

Three approaches will be considered:

- Attachment of PV cells to a fabric
- Weaving or knitting PV fibres into a fabric
- Direct coating of PV cells on to a fabric

5.2.1 ATTACHING CELLS TO A FABRIC

The first of these presents the fewest obstacles to the semiconductor fabricators but leads to a fabric that is far from the ideal flexible material. Standard silicon solar cells may be cut into smaller pieces which offer some possibilities for a deformable product by positioning these cells between folds, but with restricted overall movement. Although this approach avoids any likelihood of delamination of photovoltaic layers, it still permits cracking of brittle silicon wafers. The alternative full integration of photovoltaic constituents with a fabric should deliver the desired mechanical soundness if secure bonding is achieved.

Attaching cells to a fabric does not demand a special technical fabric and thus offers a multitude of choices to meet creative and visual criteria but does compromise

DOI: 10.1201/9781003147152-5

design by the limited choice of cell types, colours and sizes. The earliest products had to use pieces of crystalline silicon wafer cells, which are somewhat fragile and mostly dark blue or black. They could be attached by sewing into pockets or with an adhesive. As each cell only delivers a small voltage (and a current according to cell area), it was essential to connect several together, using metal conductors, often wires or ribbons, attached to the fabric. This results in a stiff framework that is only suitable for fabric that does not have to drape like normal cloth. Lightweight clothing or fashion items were rather clunky in appearance and use. Bags and backpacks were more suitable applications. A more advanced variant of attaching conventional crystalline cells to fabrics used embedded millimetre-dimensioned cells strung on copper wires and held with a liquid crystal polymer yarn inside a cast resin casing. These solar cell micropods were mounted inside knitted polyester sleeving producing solar-E-yarn that was woven into fabric. A range of tests on fabric samples in sunlight produced over 2 mW/cm^2 and demonstrated the capability of charging a battery or supercapacitor and powering some wearable devices as well as resistance to flexing and washing (Satharasinghe et al., 2019).

As thin-film cells became available, on metal or plastic foils, these extended the applications for which simple attachment of cells to fabric was acceptable. The reduction in weight and fragility was accompanied by the integration of several cells into a small PV module that delivered the required current and voltage without further interconnections. Cells and modules could even be specially shaped instead of simple rectangles and triangles that are easiest to cut from circular silicon wafers. Some fashion items were floated by designers but with a dominant metallic appearance from cells available at the time. This use also reminds us that clothing must be cleaned during its lifetime, and that any attached solar cells will have to withstand washing, which we shall return to in a later chapter. Crystalline silicon cells are naturally sealed by the hard oxide layer that rapidly forms on any native surface, but other cell types require effective hermetic encapsulation against ambient oxygen and/or water, according to their chemical makeup. If this is provided by an additional film, then it also adds to the mechanical stiffness of the product: a laminated cell attached to a fabric certainly takes us a further step from the ideal flexible PV product.

Hermetic sealing is particularly required for organic cells, many of which soon degrade when exposed to moisture or even to oxygen. One of the early strategies used by Krebs and coworkers (2006) in Denmark for incorporating this type of PV cell into a textile was simply to prepare cells on a thin PET sheet that could then be sewn as patches on to a fabric. The cell structure was built up from layers of indium tin oxide (*ITO*, a widely used transparent conductor), active polymer (MEH-PPV, Poly[2-methoxy-5-(2-ethylhexyloxy)-1,4-phenylenevinylene]) and C$_{60}$/aluminium contacts, with a second PET sheet laminated on top. The positioning of the sewn cells on a garment was dictated by the relative stiffness of the two PET sheets.

5.2.2   FORMING PV FIBRES INTO A FABRIC

In contrast, making a photovoltaic fabric without the intermediate stage of a further substrate, with or without a second protective film, would produce a device with better flexibility. One route to this solution is to coat fibres with multiple layers needed

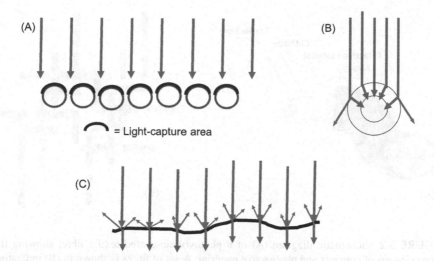

FIGURE 5.1 Light collection by fibres (A, B) and by a rough planar surface (C). The area on circular cross-section fibres that receives light is approximately 50% of the total but away from the uppermost points light may be scattered away unless it is refracted inwards (B).

for a cell, and then form these PV fibres into a fabric, for instance by knitting or weaving. Aside from the manufacturing challenge of weaving or knitting without damaging the coating, there is the topology of the long cylindrical cells to consider when connecting them together into a power module. Counter to this disadvantage is the possibility of higher light capture by fibre-shaped cells than by planar-shaped cells as they are potentially capable of omnidirectional light capture (Figure 5.1).

A scheme must be adopted for connecting to the upper and lower contacts of the many cells, because using only a single very long cell will entail too much resistance reducing the power delivered. As we have seen, connecting cells in series is always an effective tactic for reducing such losses, but it is challenging to do this for cells on a fibre that has been formed into a fabric: not only are there fibre crossovers to allow for, but also well-defined gaps will be needed between the small cells. As shown by the figure, unintended connections may exist between cells at crossovers, although these may be prevented by an insulating layer. It is more difficult to arrange for the needed connections between the top of one cell and the bottom of the next one within a manufacturing process, and another insulating layer may be needed to prevent a short circuit across the edge of each cell by the interconnecting conductor (Figure 5.2).

Several imaginative ideas have addressed these challenges, using one or more of the following supports and electrodes: metal, polymer, silicon carbide (SiC sometimes known as carborundum) or glass as wires and fibres, meshes and tapes. A specialised extension of these geometries used optical fibres having a photovoltaic coating to capture light that passed within the fibre instead of from the outside (Liu et al., 2007). Three types of cell materials have been constructed using fibres: dye-sensitized, organic and perovskite.

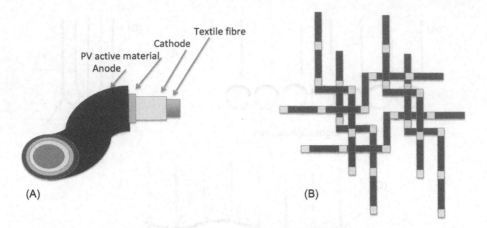

FIGURE 5.2 Schematic diagram (A) of a photovoltaic-coated textile fibre, showing the coaxial layers of contacts and photoactive material. A set of fibres is shown in (B) indicating crossovers, where there should not be any electrical contact. Intentional contact along fibres is realised by overlapping anodes and cathodes in a series chain.

It will be recalled (Chapter 2) that early DSSCs were liquid-based devices with an electrolyte between a pair of photoactive and counter electrodes. Even when this is replaced with a paste, some ingenuity is required to seal the electrolyte in place when the cells are to be integrated into a textile. As DSSCs include electrochemical steps, they are not purely photovoltaic and so are prone to degradation by more mechanisms than pure PV cells. This restriction is exacerbated by the wear and tear on fibres and coatings when they are being woven into a fabric. Early mechanical tests of fibre-shaped DSSCs examined the performance of the nanoporous TiO_2 layer on Ti wires under stress, showing that coating adhesion was the cause of electrical failure and so defining the tension that could safely be applied during textile fabrication (Ramier et al., 2008). As this method of weaving coated wire PVs into a fabric has so many technical problems caused by both frictional and tensional forces, an alternative was tried in which the upper electrode (metal ribbon coated with TiO_2/dye) and the lower electrode (platinised carbon yarn) were inserted as weft ribbons and yarns into a glass yarn textile. Additional nylon warp wires were included over this area to provide electrode separation until the gap was filled with electrolyte, completed after the whole arrangement had been laminated into a plastic pouch. This assembly could then be sewn on to a conventional fabric. Although relatively stiff, these cells could be bent over a small curved former and continued to operate with some loss of efficiency (Yun et al., 2015, 2016).

Taking a step back from forming a fabric with PV fibres, another approach to fibre-based DSSCs had used Ti wires coated with aligned TiO_2 nanotubes on which dye was absorbed and a coaxially wrapped sheet of multiwalled carbon nanotubes (MWCNT) as counter electrode. (Wrapping an MWCNT sheet electrode is an electrical improvement on a more open spiral-wrapped MWCNT fibre.) An iodine-based electrolyte was injected after the fibres had been sealed into plastic tubes. The optimised structure had a 4.1% energy conversion efficiency, was stable when flexed, and showed the expected insensitivity to the angle of incidence of the illumination but there were no reports of making these fibre DSSCs into a fabric (Sun et al., 2013). A simpler wire-based construction reported by Fu et al. (2012) had the dye-sensitized layer on titanium wires with adjacent counter-electrode stainless steel wires and the surrounding electrolyte contained within a glass or plastic envelope, giving 2.4% efficiency for a small area of $1.5\,cm^2$. More recently, an efficiency of 1.23% was achieved by using a fully organic dye in place of the more usual ruthenium ones, but with fibres up to 100 mm in length (Casadio et al., 2021). Their structure used a 250 μm Ti wire to support the dye-sensitized TiO_2 photoanode layer, with a parallel Pt counter-electrode wire, in a PTFE capillary tube containing the iodide electrolyte. An earlier paper had reported an AM1.5 10% efficient fibre-shaped DSSC (Fu et al., 2018) using a nanoparticle Pt-coated, aligned carbon nanotube electrode in a concentric hydrophobic core and hydrophilic sheath arrangement: the core provided strength and electrical conductance for the electrochemically active surface. This special electrode was wound around a Ti/TiO_2 wire photoanode that held the usual photoactive dye, and the whole ensemble was immersed in an iodide electrolyte. Five of these 30 mm long devices were successfully trialled on a T-shirt to power an adjacent pedometer. Although capable of some form of weaving, none of these are easy structures for forming into a fabric. Indeed, a review in Peng et al. (2018) charted the progress in developing fibre-shaped PVs using DSSCs and other cell types, but this concentrated on the variety of fibre structures with little on their assembly into applications. The challenges still remain severe, as detailed by Wang et al. (2020).

Perhaps one of the more successful woven DSSC fibre structures was that of Zhang et al. (2016) as this used an all-solid-state DSSC cell with two sets of interwoven wires. In one direction, a dense array of ZnO nanowires on metallised polybutylene terephthalate (PBT) fibres which were dye sensitized (using ruthenium/bis-tetrabutylammonium) and coated with CuI, and in the other direction, copper-coated counter-electrode PBT wires. Special care was taken during weaving with controlling the tension and by selecting either the photoanode or the counter electrode as the flying shuttle fibre, and the two sets of electrodes were not only connected at their crossover points but also at the edge of the textile piece. The overall efficiency was not high (1.3% at best), but the output from a small array was sufficient to demonstrate effective power for a calculator (Figure 5.3).

By replacing the DSSC structure with an organic PV, some of these construction difficulties are removed. Setting aside the degradation issues shown by some formulations of organic PV cell (OPV), whilst noting that lifetimes have not been completely studied for the DSSCs just discussed, there have been several proposals for OPVs on fibres and textiles. OPVs have shown efficiencies of at least 18% on planar substrates, somewhat higher than DSSCs but lower than perovskite PVs. Early on, it was recognised by proponents of organic active layers for PVs that despite the attractive liquid-based coating techniques, they would be difficult to lay down as even layers when fabricating them on wire-shaped substrates. A scheme for using P3HT:PCBM as the photoactive layer, within a sandwich of ITO (on standard type BFH37 multimode optical fibres) and aluminium (the outer contact), employed only fibre end-face illumination and subsequent evanescent

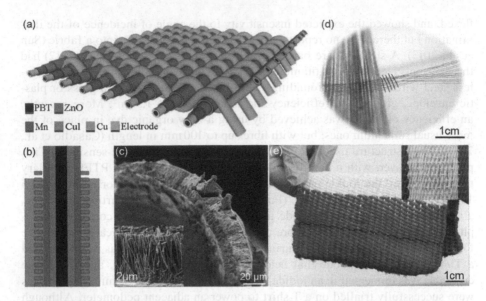

FIGURE 5.3 (a) A DSSC textile woven from metallised polymer fibre electrodes; (b) and (c) the photoanode structure and SEM image; (d) the weaving process; and (e) the PV textile mixed with woollen fibres. (From Zhang, N., Chen, J., Huang, Y., Guo, W., Yang, J., Du, J., Fan, X. and Tao, C., A wearable all-solid photovoltaic textile. *Adv. Mater.* 2016, 28, 263–269.)

optical transmission into the active layers, which did not require the dip-coated layers to have high smoothness (J. Liu et al., 2007).

Consequently, in 2009, a scheme was described by Lee et al. (2009) that used a pair of ultrasmooth stainless steel wire electrodes: the thicker of the two (100 μm diameter) supported the active coatings, and the secondary wire (50 μm), coated with silver paste, was loosely spiral-wrapped around it, before polymer coating the pair. The PV layers applied in succession, by drawing the wire through a cup of solution for each layer and drying each in an oven, were as follows: first an electron transport layer of TiO_x, then the photoactive layer of mixed P3HT and PCBM, then a hole transport layer of PEDOT:PSS. The usual current ~ voltage characteristics were obtained for lengths of 5–30 cm under 100 mW/cm^2 of AM1.5 simulated sunlight. Estimating the effective area as the product of wire length and diameter gave a best performance of 3.87% with 11.9 mA/cm^2 short circuit, 0.607 V open circuit and a fill factor of 53.8. This result was from the orientation when there was minimum shadowing of the active layers by the counter electrode, but obviously some shadowing is inevitable with this geometry for useful wire lengths. However, depending on its refractive index, the polymer cladding does partially focus light on to the active layer, and the silver coating of the counter electrode provides some diffuse reflectance around the shadowing wire, so the optics are not at all trivial. If these wire PVs had been formed into a fabric, then the optical situation would have been more complicated and light capture by textured surfaces will be discussed further in the next chapter. D. Liu et al. (2012) employed carbon nanotubes as an alternative material to this metal wire counter electrode, but their underlying structure was a polymer heterojunction on stainless steel wire and exhibited good performance when bending. 304 stainless steel wire P3HT:PCBM organic cells were also fabricated with a ZnO base contact layer and a transparent PEDOT:PSS outer electrode, all layers being dip-coated and annealed in succession (Sugino et al., 2017). A final gold wire electrode was helically wrapped around the coated wire before a final resin coating. Measured performance under standard test conditions gave estimated efficiencies up to 4.6%, taking the active area as the length times the fibre diameter. These PV wires, a few cm long, were then handloom woven into PET fabric.

Before these metal-wire-based organic PV structures, polyimide-coated silica fibres had been tried and compared directly with conventional planar organic PV devices (O'Connor et al., 2008). All layers were thermally evaporated in succession: Mg/Mg:Au/Au, copper phthalocyanine, C60, tris(8-hydroxyquinoline) Al, Mg:Ag, Ag. The performance of mm lengths of fibre cell was examined regarding the illumination incident angle, showing 0.5% power conversion efficiency for normal incidence and a more gradual decrease with zenith angle (i.e. tilt along the fibre length) than for planar cells.

An organic PV fibre that replaced the usual ITO contact layer with very thin thermally evaporated LiF/Al was based on polypropylene (PP) monofilament fibres (Bedeloglu et al., 2010). The low temperature allowed by PP restricted drying of the PEDOT:PSS lower contact and the photoactive polymer mixtures (either P3HT:PCBM or MDMO-PPV:PCBM) to 50°C. The latter of the two photoactive polymer blends performed better, giving slightly lower V_{oc} but more than twice the I_{sc} (in stc) for 50 mm lengths but even so, only 0.021% power conversion efficiency with significant series and shunt resistance losses.

Stretchability was added to the properties of flexible PV fibres by winding a Ti wire around a metal former and then removing it before fabricating it into an organic PV fibre. After coating the helical Ti wire with aligned TiO_2 nanotubes, layers of the photoactive polymer (P3HT:PCBM) and PEDOT:PSS (hole transfer layer) were added, after which a supporting elastic fibre was inserted into the coated helix, and the cell was completed by wrapping a highly aligned MWCNT sheet (Zhang et al., 2015). Charge collection was greatly enhanced by using these *aligned* nanotubes, which would otherwise be rather resistive. The power conversion efficiency was slightly reduced from a best performance of 1.23% to 1.19% after 100 cycles of stretching although under high strain the metal coil did not fully recover its shape. More significantly, fibres were woven into textile to produce clothing that was successfully tested for stretchability (Zhang et al., 2014).

Other types of fibre, such as SiC and cellulose, have been tried by various researchers but are not necessarily appropriate for textiles unless for special applications (Greulich-Weber et al., 2009; Ebner et al., 2017).

More significantly, the active material tested for fibre-shaped cells has now encompassed the promising perovskites with their large power conversion efficiency (over 21% for small area cells) and low-temperature solution processing, though still with moisture sensitivity/degradation problems. [See 'solar cell efficiency tables' for the latest perovskite PV performances (Green et al., 2022).] The rate of progress in perovskite-based fibres is evident by comparing two extensive reviews published in 2015 and 2019 (Peng et al., 2015; Balilonda et al., 2019). The earlier paper cited the novel use of MWCNTs by H. Peng's group for the outer electrode of a perovskite cell on TiO_2-coated stainless steel wire, which achieved over 3% efficiency and high bendability and anticipated further improved performance (Qiu et al., 2014). Indeed, a more recent variant of this structure on Ti wire has attained over 7% efficiency (Hu et al., 2018), but Peng's group later reached almost 10% efficiency with their improved fibre by using a flat PEN/ITO substrate strip that supported larger perovskite crystallites (Qiu et al., 2016). This report is one of very few that utilises a non-circular cross-section fibre, which can still be woven into a textile fabric.

However, the more commonly used hole-transport layer of spiro-OMeTAD [2,2', 7,7'-tetrakis(N, N-di-p-methoxy phenylamine)-9,9'-spirobifluorene] is air-sensitive, and the usual perovskite, $MAPbI_3$ [methyl ammonium lead iodide], is humidity sensitive and has potential toxicity concerns for the environment. Highly effective encapsulation is essential to combat degradation by UV, moisture, moderate temperatures and oxygen. The 2019 review of perovskite solar fibres recognises these restrictions on end-use and notes the increased attention towards lead-free perovskites, as well as suggesting that knitting would be a desirable process for fabric manufacture if the fibre structures can be better engineered and perhaps use conducting polymeric materials for the two electrical contact layers.

In conclusion to this discussion of fabricating fibre-shaped PV cells, whatever the selected photoactive material, we concur with Krebs and Hosel (2015) that the problems of electrically connecting and hermetically sealing these devices present difficulties that have not yet been fully resolved. Their tape-shaped substrate with organic PV coating overcomes the connectivity aspect by facilitating alignment of short lengths and can form woven fabrics, and as we have just seen, tapes have also been used to make efficient perovskite cells (Qiu et al., 2016). A comparison table of fibre-based solar cell performances is given by Seyedin et al. (2021).

5.2.3 DIRECT COATING OF PV CELLS ONTO A FABRIC

The third approach to making a textile solar panel is to coat the various cell layers directly on to a preformed fabric, thus avoiding the geometrical and handling problems of processing coated fibres. This is the method that retains most of a textile's fabric feel and handle but requires the coating of successive thin layers to be conformal and almost flawless on these highly structured substrates. Obviously, the first step in making any type of PV textile is to provide a conducting layer, using one of the methods that we have discussed already. Successful implementation has included one or more of the following: metallic inks, silver nanowires, transparent conducting oxides, conducting polymers and graphene. Conventional inorganic semiconductors and even DSSCs are mostly processed at high temperature and require unconventional fabrics as a base (e.g. glass fibre, metal mesh or wires), thus excluding everyday textiles.

We saw in Section 5.2.2 that DSSCs have been made with metal mesh strip anodes and carbon yarn counter electrodes that could be woven into a special glass fibre and nylon textile that allowed high-temperature sintering (Yun et al., 2015), but this was not a directly coated PV textile. In contrast, Opwis et al. (2016) also used a woven glass fibre fabric for their DSSCs, but it was coated with a smoothing layer of polyamide before adding the usual Ti/TiO_2/dye layers, an electrolyte and a transparent counter electrode with Pt clusters. Their choice of substrate enabled high-temperature curing of the TiO_2 and a final encapsulant of epoxy and plastic foil allowed operation of these $6\,cm^2$ devices for several weeks at almost 1% efficiency. The polyamide film and transparent counter electrode prevented leakage of the electrolyte, which is a potential difficulty with liquid electrolyte DSSCs. Reducing the sintering temperature of the TiO_2 is a significant improvement as it enables a wider choice of flexible substrates, especially if the usual TCO electrode is also replaced. S. Beeby's researchers have made several such improvements to textile-based DSSCs: for instance, Liu et al. (2018) used woven 65/35 polyester cotton with a levelling layer of UV-cured polyurethane, applied by screen printing, and a silver ink electrode, also screen printed, as the base. This combination withstood the 150°C sintering of screen-printed TiO_2 nanoparticles to which dye was then added although not without some deformation and delamination. The upper electrode was a platinised TCO-coated glass cover and the electrolyte was still a liquid, in this partly developed fully flexible PV. After further work, the group produced an all solid-state DSSC using the solid electrolyte Spiro-OMeTAD (N2,N2,N2′, N2′, N7,N7,N7′, N7′-octakis(4-methoxyphenyl)-9,9′-spirobi[fluorene]-2,2′, 7,7′-tetraamine) but reverting to 450°C treatment of the TiO_2 for more effective sintering (Liu et al., 2019). Such a high temperature required a different fabric than polyester cotton and a different levelling material than polyurethane, thus glass fibre textile was used, like Opwis and others (see above), with a polyimide levelling layer. The transparent upper electrode was a dual layer of PEDOT:PSS and silver nanowires. Processing of all layers, apart from the electrolyte which was drop cast, was by screen printing or spray coating, all of which are processes known to textile producers. Although the PV efficiency was only 0.4%, much less than similar devices made on glass substrates, there are trade-offs in layer thicknesses and mechanical rigidity that may yet be optimised, without moving away from all-liquid processing. The levelling layer does partly compromise the cloth-like feel of the substrate but removes the uneven texture that is so difficult to coat with multiple very thin layers (Figure 5.4).

Plentz et al. (2016) also used glass fibre fabric for their PV cells with PV-active amorphous silicon layers (a-Si:H). They also used a levelling layer of resin (butadiene styrene methacrylate) dip-coated to fill in between the fibres. The cell structure, on the weft yarns only, was TCO, a-Si:H (PIN layers) and a top contact of thin Ti, giving active areas of a few mm^2. Illumination was either through the fabric or through the Ti, with the best efficiency reaching 1.4% (in *stc*) for cells illuminated through thin Ti, limited by the optical transmission of Ti, contaminants from the resin during a-Si:H deposition and series resistance of thin Ti. Improved performance should have resulted from using a TCO in place of Ti but no results have been presented yet. As the processing temperature for plasma-enhanced chemical vapour deposition of a-Si:H is around 200°C, it is not essential to use glass fibre or other high-temperature substrate materials (Schubert & Werner, 2006), and indeed we have used woven polyester without degrading the fabric during any of the coating steps (Mather & Wilson, 2017).

As discussed previously, perovskite solar cells are in effect an extension of DSSCs but are all solid state and promise high efficiency, but still liable to ambient

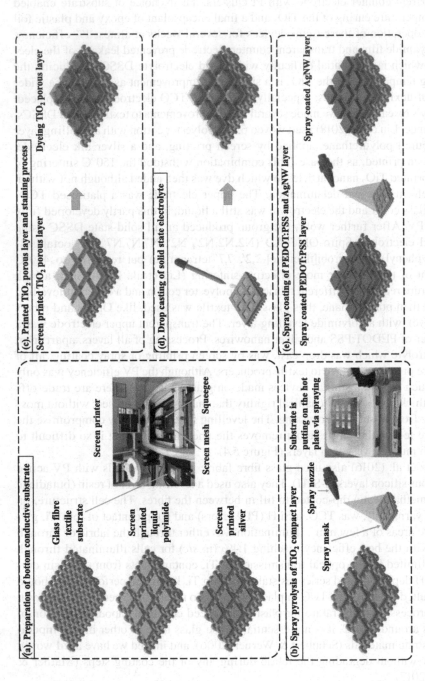

FIGURE 5.4 Solid-state DSSC on textile showing the liquid-based fabrication stages, starting with a polyimide levelling layer. (From Liu, J., Li, Y., Yong, S., Arumugam, S. and Beeby, S., Flexible Printed Monolithic-Structured Solid-State Dye Sensitized Solar Cells on Woven Glass Fibre Textile for Wearable Energy Harvesting Applications. *Sci Reports*, 2019, 9, 1362.)

degradation when not effectively encapsulated. Rapid advances in efficiency from less than 3% to over 10% were made for flexible cells on PET or PEN (polyethylene naphthalate) polymer sheet with various changes to the basic structure of spiro-OMeTAD, $MAPbI_3$, TiO_x and metal or ITO contacts, as related by Susrutha et al. (2015). Diverse fabrication techniques generally supported large-scale production possibilities and used low-temperature processing steps, but a textile perovskite device was elusive until Taiwan-based researchers reported attaching an encapsulated, polymer film-based, perovskite cell to textile that would withstand water immersion (Lam et al., 2017). Although this did not present a directly coated textile device, it demonstrated a couple of features that would be required by direct coating of a textile base: low-temperature formation of a layer of tin oxide (with the addition of a thin fullerene layer of PCBM (Phenyl-C61-Butyric-Acid-Methyl-Ester) to improve its conductivity as the cathode) and encapsulation by a 3-M commercial tape elastomer.

The feature of textiles that makes solution-based deposition rather difficult is their porosity, and the common response is to add a levelling and sealing layer of resin, as we have already seen for DSSCs. This layer may also endow hydrophilic behaviour on those fabrics that are naturally hydrophobic and that would otherwise be inimical to solution coating. Jung et al. presented a fully solution-processed perovskite textile PV which also omitted an oxide layer, based on laminating a polyurethane film to satin weave polyester fabric, which also served as a lower face encapsulant (Jung et al., 2018). Their rather small cells had a structure comprising an anode layer of PEDOT:PSS with 0.5 wt% added single-wall carbon nanotubes, then a separating layer of PEDOT:PSS, then the usual perovskite ($MAPbI_3$), then the cathode layer of PCBM, all bar-coated, with the addition of a vacuum-evaporated silver top contact and each layer processed at 100°C at most. These gave a very useful 5.72% efficiency (although note the small cell area of 6 mm^2) which remained above 80% of this value after 300 hours in ambient, showing the need for effective encapsulation. The dominant cause of a lower efficiency than for similar cells on conventional substrates was a poor fill factor, mainly due to current shunt paths across the junction.

Alternatively, one may use an organic PV compound or polymer which may also be deposited from liquid precursors and annealed at temperatures suitable for polyester or cotton. An early example from Empa, Switzerland, used a fine woven mesh of molybdenum metal wires (in one direction only) and polymer fibres, infilled on one side only with a polymer, by liquid immersion (Kylberg et al., 2011). This transparent and flexible material was the base for organic PV cells using successive layers of PEDOT:PSS (doctor blading), P3HT/PCBM (spin-coating) and opaque aluminium (vacuum evaporation) and giving efficiencies for $3–7 \text{ mm}^2$ greater than 2%. The PEDOT:PSS was thicker (>1 μm) than allowed by spin-coating (~100 nm) to provide sufficient conductivity but without blocking too much light, but once again it was the poor fill factor that restricted the performance. The woven mesh withstood bending more effectively than cells having ITO contacts, although tight bending radii would cause delamination of the polymer filling. Steim et al. (2015), also from Empa, and collaborators reported a similar organic PV using a woven mesh of metal and polymer, but in their devices the mesh was laminated on top of the completed cell (constructed on PET film) and it had no polymer filler layer. Completed cells were encapsulated between glass plates. The performance was very similar whether the cells were illuminated through the mesh side or the AZO/Ag/AZO contact side. [AZO = aluminium doped zinc oxide]. Around the same time, Lee et al. (2014) wove conducting fibres (metallised PET) into a textile electrode that was used as the anode for OPVs made with the structure ITO/ZnO/P3HT:PCBM/MoO$_3$ and stitched into normal clothing textile. Electrical contact by the OPV to the textile anode was by pressure alone, which gave resistance losses and limited the efficiency to less than 2%.

Researchers at Southampton University have made some useful advances in textile OPVs as well as in the textile DSSCs discussed above and with lower processing temperature. The first fully spray-coated textile OPVs were produced on woven 65/35 polyester cotton fabric overcoated with a UV-cured polyurethane smoothing layer (Arumugam et al., 2016). There were five layers in the OPV structure, all sprayed from appropriate suspensions in ethanol or dispersed in water or as solutions and with masking to define the active cell area of 6 mm^2:

Ag nanowires/ZnO/P3HT:ICBA/PEDOT:PSS/Ag nanowires [ICBA = indene-C60 bisadduct]. The maximum annealing temperature was 135°C, within the capability of polyester cotton. The fabric was adhered to a ceramic tile for processing and when complete was peeled off for assessment: this probably caused some micro-cracking that restricted the current density compared with similar cells on glass. In addition, the silver nanowires were sufficiently sharp and fuzzy that they could cause short circuits through the cell unless they were compressed before over-coating and thicker than desirable active layers were used. Further optimisation of layer thicknesses was anticipated to increase the low efficiency. A later paper from the same research group (Arumugam et al., 2018) still used UV-cured polyurethane (PU) to smooth the fabric surface, but this time a dual layer of hydrophobic PU to reduce swelling and a second coat of hydrophilic PU. They replaced the lower contact of silver nanowires by a spray-coated silver nanoparticle layer, and the previously sprayed ZnO nanoparticle suspension was replaced by another version that had to be doctor bladed. The improved devices gave efficiencies up to 1.23%, with low fill factors and lower short circuit currents than similar devices on glass. An encapsulating silicone resin helped to reduce bending strain during flexure tests as well as to seal the layers from ambient degradation but was not optimised. A further iteration of this key design used a different smoothing resin and a different active polymer layer (with a much greater optimum thickness), giving a lower performance than the version discussed above (especially short circuit current and fill factor), but tested the encapsulant performance further (Li et al., 2019). The encapsulation did increase the lifetime by slowing oxidation, but both tensile and compressive bending cycle tests gave a short lifetime that still requires some engineering improvement (Figure 5.5).

5.3 CONCLUSION

These three styles of producing photovoltaically active fabric each have their merits, but if the feel and behaviour of a true fabric are crucial then attaching PV cells is inadequate. The alternatives of fabricating a textile from a PV active fibre or yarn versus coating a textile with PV active layers can deliver fabric behaviour but cannot yet deliver adequate PV performance for any but the smallest energy demands.

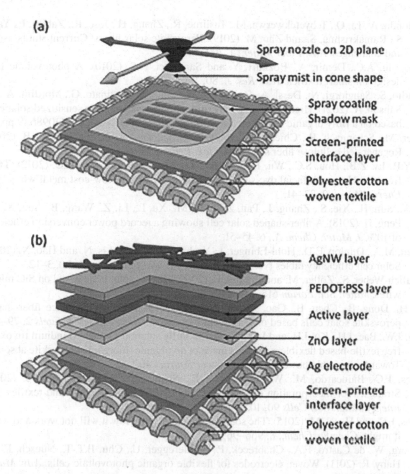

FIGURE 5.5 Spray-coated organic PV on a planarised cotton textile (first screen printed with polyurethane pastes to smooth the surface). (From Li, Y., Arumugam, S., Krishnan, C., Charlton, M.D.B. and Beeby, S.P., Encapsulated Textile Organic Solar Cells Fabricated by Spray Coating. *ChemistrySelect*, 2019, 4, 407–412.)

PV fibre still necessitates difficult engineering for electrical connections between miniature cells despite some impressive reported efficiencies for individual cells and direct coating of everyday textiles needs to overcome their roughness without excessively thick coatings. The next chapter will explore the additional characteristics that PV textiles may possess or may require to meet diverse potential applications.

REFERENCES

Arumugam, S., Li, Y., Glanc-Gostkiewicz, M., Torah, R.N. and Beeby, S.P. (2018). Solution processed organic solar cells on textiles. *IEEE J. Photovolt.*, 8, 1710–1715.

Arumugam, S., Li, Y., Senthilarasu, S., Torah, R., Kanibolotsky, A.L., Skabara, P.J. and Beeby, S.P. (2016). Fully spray-coated organic solar cells on woven polyester cotton fabric for wearable energy harvesting applications. *J. Mater. Chem. A.*, 4, 5561–5568.

Balilonda, A., Li, Q., Tebyetekwerwa, M., Tusiime, R., Zhang, H., Jose, R., Zabihi, F., Yang, S., Ramakrishna, S. and Zhu, M. (2019). Perovskite solar fibers: Current status, issues and challenges. *Adv. Fiber Mater.*, 1, 102–125.

Bedeloglu, A.C., Demir, A., Bozkurt, Y. and Sariciftci, N.S. (2010). A photovoltaic fiber design for smart textiles. *Text. Res. J.*, 80, 1065–1074.

Casadio, S., Sangiorgi, N., Dessi, A., Zani, L., Calamante, M. Reginato, G., Mordini, A. and Sanson, A. (2021). Highly efficient long thin-film fiber-shaped dye sensitized solar cells based on a fully organic sensitizer *Sol. Energy Mater. Sol. Cells* 224, 110986 (9 pp.).

Ebner, M., Schennach, R., Chien, H.-T., Mayrhofer, C., Zankel, A. and Friedel, B. (2017). Regenerated cellulose fiber solar cell. *Flex. Print. Electron.*, 2, 014002.

Fu, Y.P., Lv, Z.B., Hou, S.C., Wu, H.W., Wang, D., Zhang, C. and Zou, D.C. (2012). TCO-free, flexible, and bifacial dye-sensitized solar cell based on low-cost metal wires. *Adv. Energy Mater.*, 2, 37–41.

Fu, X., Sun, H., Xie, S., Zhang, J., Pan, Z., Liao, M., Xu, L., Li, Z., Wang, B., Sun, X. and Peng, H. (2018). A fiber-shaped solar cell showing a record power conversion efficiency of 10%. *J. Mater. Chem. A.*, 6, 45–51.

Green, M.A., Dunlop, E.D., Hohl-Ebinger, J., Yoshita, M., Kopidakis, N. and Hao, X. (2022). Solar cell efficiency tables (version 59). *Prog. Photovolt. Res. Appl.*, 30, 3–12.

Greulich-Weber, S., Zöller, M. and Friedel, B. (2009). Textile solar cells based on SiC micro-wires. *Mater. Sci. Forum* 615–617, 239–242.

Hu, H., Dong, B., Chen, B., Gao, X. and Zou, D. (2018). High performance fiber-shaped perovskite solar cells based on lead acetate precursor. *Sust. Energy & Fuels* 2, 79–84.

Jung, J.W., Bae, J.H., Ko, J.H. and Lee, W. (2018). Fully solution-processed indium tin oxide-free textile-based flexible solar cells made of an organic–inorganic perovskite absorber: Toward a wearable power source. *J. Power Sources* 402, 327–332.

Krebs, F.C., Baincardo, M., Winther-Jensen, B., Spanggard, H. and Alstrup, J. (2006). Strategies for incorporation of polymer photovoltaics into garments and textiles. *Sol. Energy Mater. Sol. Cells* 90, 1058–1067.

Krebs, F.C. and Hosel, M. (2015). The solar textile challenge: How it will not work and where it might. *ChemSusChem.*, 8, 966–969.

Kylberg, W., de Castro, F.A., Chabrecek, P., Sonderegger, U., Chu, B.T.-T., Nüesch, F. and Hany, R. (2011). Woven electrodes for flexible organic photovoltaic cells. *Adv. Mater.*, 23, 1015–1019.

Lam, J., Chen, J., Tsai, P., Hsieh, Y., Chueh, C., Tung, S. and Chen, W. (2017). A stable, efficient textile-based flexible perovskite solar cell with improved washable and deployable capabilities for wearable device applications. *RSC Adv.*, 7, 54361–54368.

Lee, M.R., Eckert, R.D., Forberich, K., Dennler, G., Brabec, C.J. and Gaudiana, R.A. (2009). Solar power wires based on organic photovoltaic materials. *Science* 324 (5924), 232–235.

Lee, S., Lee, Y. Park, J. and Choi, D. (2014). Stitchable organic photovoltaic cells with textile electrodes. *Nano Energy* 9, 88–93.

Li, Y., Arumugam, S., Krishnan, C., Charlton, M.D.B. and Beeby, S.P. (2019). Encapsulated textile organic solar cells fabricated by spray coating. *ChemistrySelect* 4, 407–412.

Liu, J., Li, Y. Arumugam, S., Tudor, J. and Beeby, S. (2018). Screen printed Dye-Sensitized Solar Cells (DSSCs) on woven polyester cotton fabric for wearable energy harvesting applications. *Mater. Today: Proc.*, 5, 13753–13758.

Liu, J., Li, Y., Yong, S., Arumugam, S. and Beeby, S. (2019). Flexible printed monolithic-structured solid-state dye sensitized solar cells on woven glass fibre textile for wearable energy harvesting applications. *Sci. Reports* 9, 1362.

Liu, J., Namboothiry, A.G. and Carroll, D.L. (2007). Fiber-based architectures for organic photovoltaics. *Appl. Phys. Lett.*, 90, 063501 (3 pp.).

Liu, D., Zhao, M., Li, Y., Bain, Z., Zhang, L., Shang, Y., Xia, X., Zhang, S., Yun, D., Liu, Z., Cao, A. and Huang, C. (2012). Solid-state, polymer-based fiber solar cells with carbon nanotube electrodes. *ACS Nano* 6, 11027–11034.

Mather, R. and Wilson, J. (2017). Fabrication of photovoltaic textiles. *Coatings* 7, 63 (21 pp.).

O'Connor, B., Pipe, K.P. and Shtein, M. (2008). Fiber based organic photovoltaic devices. *Appl. Phys. Lett.*, 92, 193306 (4 pp.).

Opwis, K., Gutmann, J.S., Alonso, A.R.L., Henche, M.J.R., Mayo, M.E., Breuil, F., Leonardi, E. and Sorbello, L. (2016). Preparation of a textile-based dye-sensitized solar cell. *Int. J. Photoenergy* 2016, Article ID 37960741 (11 pp).

Peng, M., Dong, B. and Zou, D. (2018). Three dimensional photovoltaic fibers for wearable energy harvesting and conversion. *J Energy Chem.* 27, 611–621.

Peng, M. and Zou, D. (2015). Flexible fiber/wire shaped solar cells in progress: Properties, materials, and designs. *J. Mater. Chem. A* 3, 20435 (24pp).

Plentz, J., Andra, G., Pliewischkies, T., Bruckner, U., Eisenhawer, B. and Falk, F. (2016). Amorphous silicon thin-film solar cells on glass fiber textiles. *Mater. Sci. Eng. B* 204, 34–37.

Qiu, L., Deng, J., Lu, X., Yang, Z. and Peng, H. (2014). Integrating perovskite solar cells into a flexible fiber. *Angew. Chem., Int. Ed.*, 53, 10425–10428.

Qiu, L., He, S., Yang, J., Jin, F., Deng, J., Sun, H., Cheng, X., Guan, G., Sun, X., Zhao, H. and Peng, H. (2016). An all-solid-state fiber-type solar cell achieving 9.49% efficiency. *J Mat Chem A* 4, 10105–10109.

Ramier, J., Plummer, C.J.G., Leterrier, Y., Manson, J.-A.E., Eckert, B. and Gaudiana, R. (2008). Mechanical integrity of dye-sensitized photovoltaic fibers. *Renew. Energy* 33, 314–319.

Satharasinghe, A., Hughes-Riley, T. and Dias, T. (2020). An investigation of a wash-durable solar energy harvesting textile. *Prog. Photovolt. Res. Appl.*, 28, 578–592.

Schubert, M.B. and Werner, J.H. (2006). Flexible solar cells for clothing. *Mater. Today* 9, 42–50.

Seyedin, S., Carey, T., Arbab, A., Eskandarian, L., Bohm, S., Kim, J.M. and Torrisi, F. (2021). Fibre electronics: Towards scaled-up manufacturing of integrated e-textile systems. *Nanoscale*, 13, 12818–12847.

Steim, R., Chabrecek, P., Sonderegger, U., Kindle-Hasse, B., Siefert, W., Kroyer, T., Reinecke, P., Lanz, T., Geiger, T. Hany, R. and Nuesch, F. (2015). Laminated fabric as top electrode for organic photovoltaics. *App. Phys. Lett.* 106, 193301 (4 pp.).

Sugino, K., Ikeda, Y., Yonezawa, S., Gennaka, S., Kimura, M., Fukawa, T., Inagaki, S., Konosu, Y., Tanioka A. and Matsdumoto, H. (2017). Development of fiber and textile-shaped organic solar cells for smart textiles. *J. Fiber Sci. Technol.* 73, 336–342.

Sun, H., You, X., Yang, Z., Deng, J. and Peng, H. (2013). Winding ultrathin, transparent, and electrically conductive carbon nanotube sheets into high-performance fiber-shaped dye-sensitized solar cells. *J. Mater. Chem. A.*, 1, 12422 (4 pp).

Susrutha, B., Giribabu, L. and Singh, S.P. (2015) Recent advances in flexible perovskite solar cells. *Chem. Commun.* 51, 14696–14707.

Wang, L., Fu, X., He, J., Shi, X., Chen, T., Chen, P., Wang, B. and Peng, H. (2020). Application challenges in fiber and textile electronics. *Adv. Mater.* 32, 1901971 (25 pp).

Yun, M.J., Cha, S.I., Kim, H.S., Seo, S.H. and Lee, D.Y. (2016). Monolithic-structured single-layered textile-based dye-sensitized solar cells. *Sci. Rep.* 6, 34249 (8 pp).

Yun, M.J.; Cha, S.I.; Seo, S.H.; Kim, H.S.; Lee, D.Y. (2015). Insertion of dye-sensitized solar cells in textiles using a conventional weaving process. *Sci. Rep.* 5, 11022 (10 pp).

Zhang, Z., Yang, Z., Deng, J., Zhang, Y., Guan, G. and Peng, H. (2015). Stretchable polymer solar cell fibers. *Small* 11, 675–680.

Zhang, N., Chen, J., Hunag, Y., Guo, W., Yang, J., Du, J., Fan, X. and Tao, C. (2016). A wearable all-solid photovoltaic textile. *Adv. Mater.* 28, 263–269.

Zhang, Z., Yang, Z., Wu, Z., Guan, G., Pan, S., Zhang, Y., Li, H., Deng, J., Sun, B. and Peng, H. (2014). Weaving efficient polymer solar cell wires into flexible power textiles. *Adv. Energy Mater.*, 4, 1–6.

6 Technological and Design Specifications

6.1 INTRODUCTION

We have concentrated on the electrical properties of PV materials without much consideration of where and how PV textiles may be employed. Each user will have a set of additional requirements that should be met, such as the combination of voltage and current required by the electrical load, type of electrical connections, fabric shape, water resistance and washability, flexure and bendability and so on. Integration of electrical storage batteries and active electronics such as miniature sensors may also be specified. Furthermore, textile substrates have more complicated optical behaviour than simple smooth flat surfaces, which may make it more difficult or perhaps easier to collect sunlight. Even then, the illumination may come from artificial lighting for indoor applications, for which there are no standard test conditions. This chapter addresses some of these concerns about operating conditions but may raise even more questions!

6.2 OPTICAL ABSORPTION

Conventional rigid solar panels usually have a glass plate cover, which is the first material presented to the illumination and geometric optics describe how light is reflected at the two glass surfaces depending on incident angle and the optical density ('refractive index'). This reflectance loss may be reduced, at some cost, by adding a tailored layer of antireflection material or by carefully roughening the external surface of the glass. The reflectance loss from a glass surface is only a few percent of incident sunlight but increases significantly for light at shallow angles to the surface, such as early morning or evening for open-air panels. Solar panels on plastic film with plastic encapsulant will behave in a similar manner.

The 'careful roughening' mentioned above as an alternative to antireflection coatings will scatter light rather than specularly reflecting it (which would produce an image of the light source), with both forward and backward reflection according to the shapes of the fine asperities. (A dusty solar panel will also scatter light as well as partially shading it.) A solar panel on a textile surface will behave more like this than a smooth surface so it is relevant to ask if such optical scatter can be beneficial. Indeed, it is possible to enhance the scattering of light by using extremely fine repetitive patterns on a surface, with the effect described by 'physical optics' rather than the ray tracing of geometric optics. The procedure explains the origin of optical interference and diffraction but requires patterning on the scale of optical wavelengths, namely nanometres rather than microns in precision. Similar interference effects are observed for thin transparent coatings of different refractive index

DOI: 10.1201/9781003147152-6

from the supporting material, such as the coloured patterns seen from oil slicks. A more complicated enhancement is by nanometre particles of some metals that provide 'plasmon resonance' on their surfaces depending on the particle size and shape (Figure 6.1). This effect is a collective oscillation of conducting electrons in the metal, and essentially these particles act as tiny antenna for a limited wavelength band of incoming visible or near-infrared light, leading to energy transfer into the adjacent semiconductor (e.g. Tvingstedt et al., 2007).

Several of these methods have been used to enhance optical absorption in rigid solar panels, but with extra cost to offset gains. The efficiency of thin-film solar cells is particularly sensitive to their thickness: too thick and charges will not be separated by the limited-extent electric field; too thin and some light will pass through the cell without complete absorption. In thin cells, optical scattering at the illuminated surface may significantly enhance optical absorption by producing longer oblique light paths through the cell.

The science behind light trapping in solar cells is explained by Kowalczewski et al. (2013) for thin crystalline silicon cells showing the difficult task of making this work for a wide band of wavelengths and revealing the advantage of a certain amount of random roughness. Rough surfaces are described quantitatively by the height deviation and average spacing of the asperities which reveals that reflection

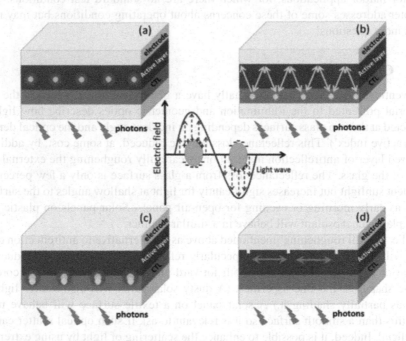

FIGURE 6.1 Plasmonic enhancement of optical absorption in a solar cell by incorporating metal nanoparticles (e.g. silver or gold) of tailored shape and size. These may lie within the active layer (a), or the charge transport layer (CTL), (b) and (c) or at the metal contact interface (d). However, they may also trap mobile charges unless each particle is covered with an inorganic shell. (From Feng, L., Niu, M., Wen, Z. and Hao, X., Recent Advances of Plasmonic Organic Solar Cells: Photophysical Investigations. *Polymers*, 2018, 10, 123 (33 pp).)

is reduced for crystal silicon surfaces having greater height variation, especially for short-wavelength light, which translates into higher current density from PV cells with this roughness of surface. When light is scattered into larger angles into the material ('haze'), there is a longer path for absorption to occur, reducing the proportion of light that would otherwise be transmitted through thin materials, especially at longer wavelengths for which the optical absorption is weaker. Light trapping has been measured by reflectance for thin-film silicon cells on textured ZnO:Al (conducting zinc oxide: aluminium) overcoated with silver, on smooth glass, for which the texturing may be controlled (Bittkau et al., 2012). The scattering at this rear contact of the cell was greater than for front surface scattering, which leads to more efficient light absorption in the silicon for longer wavelengths. There have been no such detailed studies for textile PV cells, thus the optimum texture has not been predicted and the wide variety of textile structures has yet to be explored. Furthermore, we have not considered non-woven fabrics as substrates for PV cells, and these materials have strong random optical scattering that may be characterised by such techniques as optical Fourier transform measurements, if these fabrics are suitably stable for electrical continuity.

Not all thin-film PV cells benefit from optical scattering: the organic PVs with textile electrodes constructed by Steim et al. (see Chapter 5) had the same short circuit current density for illumination through a hazy fabric electrode as when illuminated through the smooth PET substrate, according to software simulation. In contrast, individual, miniature, fibre-shaped DSSCs are able to capture light from all directions. When illuminated from one direction but with a diffuse reflector placed beneath, there was an increase in short circuit current of more than twice that when perpendicular illumination was only one-sided (Fu et al., 2011). For organic PVs on a fabric woven from bundles of 10 μm fibres, the photocurrent is definitely enhanced by the scattering of incident light from the curved surface of individual fibres compared with a similar conventional flat substrate PV cell (Lee et al., 2014). The practical disadvantage of very rough surfaces is their often poorer electrical behaviour, and indeed many of the textile PVs have poorly shaped current versus voltage characteristics (i.e. low fill factors) that indicate underlying resistance losses.

A different optical enhancement is being used with some traditional PV cells to increase the absorption of light from wavelengths outside the well-absorbed band. They use a luminescent dye in a polymer matrix to convert some incoming light to another wavelength that is more readily absorbed, giving enhanced photocurrents and a colour option: these can convert high energy (blue) light to lower energies or, less obviously, rare-earth ions such as lanthanides can convert infrared to visible. Recently the concept has been tested with fibre-shaped DSSCs in which the dye is held in a plastic sheet adjacent to the array of fibre cells. Light is thus captured over a bigger area than that of the fibres alone and is re-emitted from the luminescent collector towards the fibre cells, according to the geometry employed (Huang et al., 2021; Figure 6.2).

Most solar cells have a simple response to increasing or decreasing light intensity: the photocurrent is proportional to light intensity, as long as the light is within the absorption wavelengths of the material. In a few types of semiconductor cells, such as amorphous silicon, the light may also change the electrical resistance of the cell

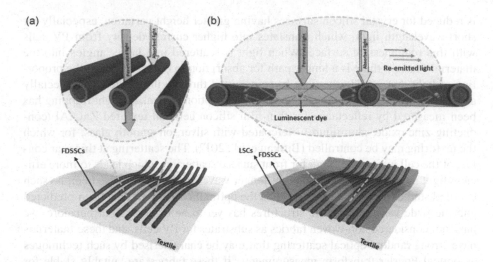

FIGURE 6.2 Luminescent solar concentrator enhancement of optical absorption by fibre-shaped DSSCs. (a) Shows that light will pass through the gaps between fibres laid on a textile backing. (b) Shows that the addition of luminescent dye between the fibres can convert incident light to another colour and guide it sideways towards the fibre DSSCs, thus using the whole textile area productively. (From Huang, C., Kang, X., Rossi, R.M., Kovalenko, M.V., Sun, X., Peng, H. and Boesel, L.F., Energy harvesting textiles: using wearable luminescent solar concentrators to improve the efficiency of fiber solar cells. *J Mat Chem A,* 2021, 9, 25974-25981.)

to an appreciable extent. This shows in the current versus voltage characteristic as a change in the slope of the curve near the open-circuit voltage region. Consequently, the performance of these types of cells has a more complicated dependence on the illumination received for indoor operation under room lighting.

6.3   TEMPERATURE EFFECTS

In Chapter 2, the standard test conditions (stc) were introduced as a worldwide set of measurement parameters that allowed the comparison of different solar PV cells and panels. These established the intensity and spectrum of the light source and the PV operating temperature, in a combination that might not be met in reality, but that could be provided by a solar simulator lamp and cooling in any location. Textile PV devices may not always be intended for outdoor use, and standard solar illumination will not then be an appropriate specification for electrical output. Indoor illumination will differ in both spectrum and intensity from outdoor sunlight. Note that the standard testing temperature for PV is 25°C, different from the 20°C standard atmosphere specified for testing of textiles. In addition, the ambient temperature will have a wide range and this alone will affect the electrical output of any PV device, as mentioned in Section 2.4. Generally, the higher the temperature the lower the output voltage. Wider bandgap semiconductors (those that have a shorter cut-off wavelength for absorbed light) are less susceptible to these temperature penalties, and some, therefore, work more effectively indoors than traditional silicon. It is not practical to actively cool most solar PV installations and so their output does decrease in bright sunlight. Textile substrates will have similar restrictions, and this must be considered when specifying the size and number of cells for any application.

6.4   MECHANICAL CONSIDERATIONS

Although structural deformation of textile devices may be considered a problem to be overcome, this will be a requirement for some sensing applications. We have mentioned only a few of the tests that PV textiles have endured such as bending but stretching and twisting may be expected during some uses. There are well-established and internationally defined procedures for testing textile reactions to different stresses. Standard tests for electrically active textiles are less widespread (e.g. BS EN 16812:2016 describes a method for measuring the electrical resistance of conducting tracks), but a variety of bending tests has been reported, such as those by Lee et al. (2014) for their textile electrode (designed for supporting OPV cells and stitched to another textile) which showed recovery of the initial conductance after multiple bending cycles. However, thermo-mechanical stress is well known to degrade the electrical conductivity of connecting tracks and bonded joints in flexible electronics, to the extent of complete failure in the most extreme conditions.

A technique for providing stretchable connections is to form them in a serpentine shape, but this does not resist bending very well. Truly elastic conductors are better but often use metal nanowires or carbon nanotubes in their formulation, and conducting polymers are much more resistive than metals and do not meet photovoltaic requirements. Good mechanical design can increase the endurance of flexible electrical interconnections when bent, by considering the position of the neutral axis in a sandwich of materials: one side of this will be compressed and the opposite will be expanded when the pair is bent, so positioning critical connections on that axis will expose them to less stress. This option is not available for the bonded joints that must be present when connecting components to a flexible circuit, such as a textile PV array feeding an external load. These points are abrupt interfaces between different materials of different thicknesses, perhaps metal tracks and metal epoxy pastes, and are known to crack when contorted. There is not yet a good solution to this problem.

6.5   WASHABILITY

One of the first demands of potential users of flexible electronics on textiles is 'are they washable?'. There are stringent tests for clothing fabrics (e.g. ISO 63330 for domestic washing and drying as well as those by ASTM) and more severe conditions for industrial applications, all of which offer difficult requirements for smart textiles, including those with integrated PV arrays. Only a few of the PV textile examples that we have described have been put under any type of washing test, and it is obvious that conventional rigid cells simply attached to a fabric are likely to be damaged by machine washing. However, by encapsulating millimetre-sized crystalline silicon cells in individual resin cylinders, which were linked by copper wires, a more robust woven fabric was produced (see Chapter 5). Patches of these fabrics, which generated 43 mW in one sun illumination, were machine washed in detergent and dried for 25 cycles and compared with another set that was hand washed for 25 cycles. All the hand-washed samples functioned and 60% of the PV yarns functioned correctly after the machine wash cycles, but there were 10% and 13% reductions in output power respectively, probably due to increased shading by the other fibres in the fabric (Satharasinghe et al., 2020).

A few researchers have applied less severe washing trials to their own variety of flexible OPV, including some organic cells on thin films of parylene (different forms of poly-para-xylylenes with substitutions for some hydrogen atoms), coated on both sides with an acrylic elastomer, showing that even this type of cell that is very sensitive to moisture can be protected (Jinno et al., 2017). A more complicated inorganic/polymer composite encapsulant also worked well in preventing the degradation of OPVs during successive bending and washing trials (Jeong et al., 2019). Although an effective atmospheric barrier for OPVs is provided by atomic layer epitaxy of multiple layers of Al_2O_3 and ZnO, this degrades in water, so the addition of a PVA/SiO_2 resin composite was introduced to protect the Al_2O_3. The OPVs in these trials were formed on PEN textile substrates with an active layer of PTB7-Th:PC71BM between electron and hole transport layers, with silver electrodes. The multi-layer encapsulant was applied directly to the textile before forming the OPVs and then a second encapsulant was attached on top. The encapsulation is compatible with roll-to-roll manufacture and successfully maintained the polymer PV output after 20 washes (Figure 6.3).

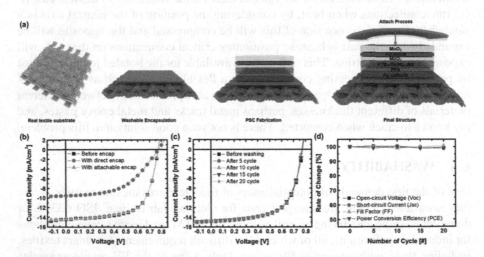

FIGURE 6.3 Encapsulated polymer PV on PEN textile showing washability. (From Jeong, E.G., Jeon, Y., Cho, S.K. and Choi, K.C. Textile-based washable polymer solar cells for opto-electronic modules: toward self-powered smart clothing. *Energy Environ. Sci.*, 2019. 12, 1878-89.)

6.6 SAFETY ASPECTS

The preceding sections have discussed environmental and operating conditions that can impinge on immediate performance as well as on the durability of PV textiles. The lifetime of these power sources is presently uncertain as none have been in use for the extended periods over which conventional PV panels have operated. Certainly, conventional solar PV panels have expected lifetimes of 25 years or more, with an increasing understanding of the few degradation effects that have been exposed over the long term. In general, rapid ageing is not an intrinsic property of PV cells but can result from poor technology (semiconductor defects that trap charges at some unintended impurities) or poor assembly into panels (e.g. ineffective encapsulant that enables ingress of moisture to attack metal tracks). By contrast, textile PV cells may have new degradation mechanisms arising from the texture and flexure of the material. However, the desired lifetime of a PV source depends on the particular application and its anticipated usage, which will be more obvious in Chapter 8.

Other considerations from the user's point of view are safety related, both during use and at end-of-life disposal. This issue is already a concern for plastic products throughout the world, particularly when they are discarded. Electrical and electronic products are not as easily thrown away: since the introduction of the European Directive on Waste Electrical and Electronic Equipment (WEEE) in the early 2000s, these goods must be collected and safely recycled. This treatment recovers materials that are valuable and may be toxic if they enter the biosphere. Recycling of solar panels will be discussed in the next chapter together with manufacturing that will allow effective recovery of the embedded materials. Hence it is most undesirable to include any toxic materials in PV construction even if the potential for harm could only arise at end of life.

However, there are some photovoltaic semiconductors that contain toxic elements, such as lead, cadmium, tellurium and selenium, and even silicon usually contains parts per million of intentional dopants that in larger amounts would be hazardous, such as phosphorus or arsenic. No-one should be anxious about these ingredients when in normal everyday use (after all, we do not worry about highly reactive sodium and toxic chlorine when they are safely bound together as table salt). Any anxiety should be reserved for disastrous situations such as a fire when they might be fragmented and decompose. Thus, lead is a cumulative poison with no lower limit for exposure, and its inclusion in the most efficient perovskite PV cells is presently a factor in reducing their commercial attraction, as it provides a risk if these cells were damaged and water attacked them. Similarly, CdTe PV cells are an efficient thin-film alternative to crystalline Si cells but are composed of two toxic elements that are safe when bound together but potentially hazardous if separated. CIGS cells have similar concerns, but like the other compound semiconductor PV options, only present a risk when disassembled. It is apparent that the perceived risks are highest at end-of-life disposal when the safe recovery of all materials should be undertaken (e.g. International Energy Agency Photovoltaic Power Systems Programme, Task Force 12 on sustainability: https://iea-pvps.org/research-tasks/pv-sustainability/). Textile-based PV should be no different in risk to conventional glass-based PV panels.

Fire has just been mentioned as presenting a risk by dispersing PV constituents, but PV installations on roofs do not present an intrinsic fire risk to the buildings beneath. The addition of an electricity-generating array must not provide a fire hazard, but it may do so if installation regulations are not complied with, such as mismatched connectors or undersized cables. These types of failings have been the cause of some much-publicised fires in countries having less regulated installation guidance than in Europe. The expected uses of PV textiles are likely to have smaller current capacities than today's rooftop arrays which greatly reduce the possibility of ohmic heating in resistive components, such as small cross-section cables or corroded contacts and junction boxes. The flammability of textile substrates is well understood and depends on their composition and on the conditions such as the available oxygen. Flame retardancy may be imparted by special coatings, but there does not appear to be any experience of combining such treatments with any of the PV coatings. The PV coatings themselves are not flammable materials.

6.7 CONCLUSIONS

When designing a PV textile application, there needs to be consideration of external operating conditions such as temperature and mechanical stress. These factors are in addition to ensuring sufficient light collection to deliver the required electrical output. Textiles will have different responses from conventional rigid substrates but may have advantages in collecting more light from diffuse sources. Additional factors may include withstanding immersion in water, especially during washing, which has been confirmed for some PV textiles. The influence of safety issues, such as flammability and toxicity of constituents, is always important but using textile substrates instead of rigid materials should not introduce new risks.

REFERENCES

Bittkau, K., Bottler, W., Ermes, M., Smirnov, V. and Finger, F. (2012). Light scattering at textured back contacts for n-i-p thin-film silicon solar cells. *J. Appl. Phys.*, 111, 083101.

Fu, Y., Lv, Z., Hou, S., Wu, H., Wang, D., Zhang, C., Chu, Z., Cai, X., Fan, X., Wang, Z.L. and Zou, D. (2011). Conjunction of fiber solar cells with groovy micro-reflectors as highly efficient energy harvesters. *Energy Environ. Sci.*, 4, 3379–83.

Huang, C., Kang, X., Rossi, R.M., Kovalenko, M.V., Sun, X., Peng, H. and Boesel, L.F. (2021). Energy harvesting textiles: using wearable luminescent solar concentrators to improve the efficiency of fiber solar cells. *J. Mat. Chem. A.*, 9, 25974–25981.

Jeong, E.G., Jeon, Y., Cho, S.K. and Choi, K.C. (2019). Textile-based washable polymer solar cells for optoelectronic modules: toward self-powered smart clothing. *Energy Environ. Sci.*, 12, 1878–89.

Jinno, H., Fukuda, K., Xu, X., Park, S., Suzuki, Y., Koizumi, M., Yokota, T., Osaka, I., Takimiya, K. and Someya, T. (2017). Stretchable and waterproof elastomer-coated organic photovoltaics for washable electronic textile applications. *Nat. Energy* 2, 780–785.

Kowalczewski, P., Liscidini, M. and Andreani, L.C. (2013). Light trapping in thin-film solar cells with randomly rough and hybrid textures. *Opt. Express* 21, DOI: 10.1364/OE.21.00A808 (13 pp.).

Lee, S., Lee, Y., Park, J. and Choi D. (2014). Stitchable organic photovoltaic cells with textile electrodes. *Nano Energy* 9, 88–93.

Satharasinghe, A., Hughes-Riley, T. and Dias, T. (2020). An investigation of a wash-durable solar energy harvesting textile. *Prog. Photovolt. Res. Appl.*, 28, 578–592.

Tvingstedt, K., Oersson, N. and Inganas, O. (2007). Surface plasmon increase absorption in polymer photovoltaic cells. *Appl. Phys. Lett.*, 91, 113514.

7 Manufacturing
Moving from Laboratory to Production

7.1 INTRODUCTION

Much of the previous content has been about research and not manufacturing, at any scale, as solar textiles are mostly at a research or early development stage. Nonetheless, production methods that have been used for other types of solar panels or for textile products may be applicable after some modifications. Some small-scale production equipment is commercially available for flexible solar cells on plastic film but will require more work before textiles are coated routinely. Some of the work cited in the previous chapters has included possible routes to commercial production and has offered alternative coating procedures that are more amenable to regular fabrication rather than single-sheet, laboratory, batch processing. In research, various PV cell layers have been laid down by both chemical and physical techniques, with either liquid or gaseous precursors and sometimes from vapourised solids. These include electroless and electro-plating, liquid (ink) printing, dip-coating, vacuum evaporation, atomic layer deposition, sputtering, chemical vapour deposition and polymerisation.

Several of these are more familiar to the electronics sector than to textiles, but the textile industry is very accustomed to liquid coating and printing on fabric rolls and there are moves towards some gas-based processing, such as plasma treatments, to avoid treating large quantities of liquid waste. Plasma chemical vapour deposition (PECVD) has been used for many years to coat thin-film amorphous silicon on to a variety of substrates including flexible polymer sheet but requires toxic and flammable gases which are again more familiar to microelectronics manufacturers than to textile producers. Uni-Solar Inc., one of the pioneers of this technology for thin-film solar panels, was established by Energy Conversion Devices in the early days of amorphous silicon research and continues to manufacture triple-junction cells on a flexible support in a roll-to-roll (R2R) process, not yet adapted for direct coating on textiles. Over the course of 25 years, they moved from laboratory 2×2 inch^2 substrates to 15×14 inch^2 sheets, to full roll-to-roll equipment. This planned development required a technology roadmap to identify the limitations and the necessary enhancements to achieve targeted cost/performance figures (e.g. grid parity). We are not aware of an equivalent roadmap for textile PV but there has been a series of International Technology Roadmaps for PV (ITRPV) with the 12th edition in 2021 that mainly refers to crystalline silicon cells and their implementation in panels. Very recently, HyET Solar, a company in the Netherlands, has begun large-scale roll-to-roll coating of amorphous and nanocrystalline silicon on aluminium foil which is

DOI: 10.1201/9781003147152-7

69

then encapsulated between polymer sheets to give 'Powerfoil', a flexible array of 12% efficient cells. A recent review of textile envelope photovoltaics for building integration gives a brief introduction to R2R coating, lamination and encapsulation that will be needed for large-scale commercial production (Li & Zanelli, 2021).

Solar photovoltaics are usually presented as renewable or sustainable energy sources, which means that they must deliver more energy during their lifetime than was used in their manufacture, a requirement that is met if they do not rapidly degrade and are not deployed in a weak solar climate. Other environmental concerns are that they should not emit large quantities of greenhouse gases during manufacture and use, and that they should be recyclable when they are no longer usable. These points are considered here after methods for manufacturing flexible PV have been reviewed.

7.2   CHALLENGES FOR TEXTILE COATING

It will be realised that the very flexible nature of textiles, which gives fabrics their particular feel compared with less flexible polymer sheet, presents coating equipment designers with a two-dimensional instability. As well as tensioning the material along its length, it must be free of wrinkles across its width and not wander between supports if it unwinds from a feed roller to a take-up roller over intermediate ones. Fabrics also tend to have a greater moisture content than plastic sheet which will be a problem for vacuum-based processes, whether or not the process requires a heated substrate, and will necessitate an outgassing step before coating. Both plastic and fabric substrate materials are more difficult to heat uniformly than a ceramic or metal sheet, and when heated, they will expand and be more easily stretched. Stresses that arise during heating or from deposited layers with differing thermal expansions all conspire to alter the planarity of flexible substrates.

Some fabric materials will readily shed particulates, with subsequent defects in any coating wherever the particles are redeposited, and possibly affecting the coating procedure itself as particles drift through the local environment. The semiconductor and electronics industries are familiar with strategies to reduce the influence of particulates, including highly filtered clean rooms for critical processes, albeit at high operational cost. There will be a reduced need for the highest-rated clean-rooms if substrates are contained within a clean enclosure during all fabrication stages, for instance in a roll-to-roll coater that provides all the steps without exposing the substrate roll to air.

Another aspect to design into any process sequence is the requirement for defining the area to be coated, which will differ between layers and will require moveable masking. Although photolithography, so familiar to microelectronics fabrication, has been incorporated into roll-to-roll fabrication by stepping and repeating the pattern in successive areas of the substrate, it is overcomplicated for simple PV devices. This is a subtractive technique in which a photosensitive film is sprayed or spun on and dried, then some areas are masked and exposed to UV light, with subsequent removal of softened areas giving a pattern, through which the underlying material can be etched. These steps are then repeated for each layer of the device. Alternatively, a shadow mask may be used to select areas to be coated, either in contact with the substrate or spaced a short distance away. Line-of-sight coating, such as vacuum evaporation, may use out of contact masking, but sputtering and PECVD will send the coating material around mask apertures to deposit only poorly defined patterns. An early arrangement for producing flexible thin-film amorphous silicon cells in a step-and-repeat procedure used sprung wire masks that could be lifted as the substrate web was moved and then replaced on the surface during static PECVD coating, without abrading the surface which is a snag with contact masks (Schubert & Merz, 2009). Textile substrates have a rough topography that will not support fine patterns in an overcoating layer and that will make it difficult to register one layer accurately over another unless wider margins are used. Alternatively, roller printing may be used with liquids in various gravure or imprint methods.

Laser machining is widely used to cut silicon wafers and to open holes for back contacts in state-of-the-art crystalline silicon PV cells. PV amorphous silicon cells have been patterned after silicon deposition, by using laser ablation to define the cell area by removing a track of silicon around the edge of the area without burning into the substrate, and laser machining systems are now sold for high-speed R2R processing of flexible organic PV cells.

The layers that form all PV devices must be laid down sequentially from precursors, noting that liquids do not require the pumped chambers needed by gases. However, commonly used methods for producing traditional PV cells on crystalline silicon wafers are mostly vacuum based, requiring separate process chambers for each step. To avoid too much exposure to air between steps, manufacturers use cluster tools with separate, gated, process chambers and a central server chamber by which wafers are moved between each process, providing semi-continuous manufacturing for moderate-sized single substrates (Figure 7.1).

Similar systems are also used to coat glass plates for liquid crystal displays. This arrangement can be adapted to handle a small roll of flexible substrate by having it spooled on a reel that is moved from chamber to chamber and then unrolled across each coating source in turn. However, for longer or wider rolls, the material has either

FIGURE 7.1 Large-area PECVD amorphous silicon coater for liquid crystal displays, by permission of Applied Materials Inc.

to be moved through the sequence of coating sources within a single large chamber, which contains both feed and take-up rollers, or moved along a linear set of chambers with differential pumping and slit valves between them as noted above. In all R2R systems, to keep the roll moving at a constant speed, chambers must be repeated or extended so that slower processes (i.e. either slower coating rate or greater coating thickness) are compatible with faster ones.

7.3 COATING OPTIONS FOR FLEXIBLE PV

An early integration of solar cells and electrical storage on cotton textile by roll-to-roll assembly was reported in Gao et al. (2016). This offered a production process for combining supercapacitor electrical storage with solar cell electrical generation into a hybrid flexible power pack. The focus of the report was on fabricating the special metallic-layered double hydroxide supercapacitors on activated cotton textile having a graphene conductive coating, mainly by wet processing and thermal annealing. The combination of commercially sourced, flexible, thin-film solar cells with these supercapacitors provided energy to light an LED for some time, whether or not the cells were illuminated. The combination was formed by a facile roll-to-roll step using double-sided adhesive tapes to join the pieces together as they passed through small-sized (four-inch wide) steel rollers. Before this could be done, the separate layers of each device would have to be fabricated as indicated above, using appropriate coating techniques. On a larger scale, the Solar Cloth company in France uses CIGS photovoltaic cells made by high-temperature vacuum evaporation on to a plastic film, then laminating films of ethylene tetrafluorethylene on both sides, giving a sandwich that can be bonded on to a woven material such as Dacron or on to sails for large yachts.

Examining the various components in PV arrays, we can start with the electrical contacts. There are several techniques that have been used on a large scale for rapidly metallising plastic or paper. Although metal inks are used to screen-print patterned conductors, their relative cost and complex chemistry may make vacuum evaporation or sputtering more amenable for large-area contacts where fine definition is not required. Sputtering is also used on a large scale for transparent conducting oxides such as ITO for touchscreens and thin-film PV at metres per minute speeds. Commercial roll-to-roll equipment such as that from Applied Materials offers multiple sources in a single chamber for sputtered or evaporated metal layers (Figure 7.2). Packaging is already metallised at speeds in excess of 10 m/s (Figure 7.3).

The active semiconductor layers require greater control of the environment around the coating process as most are sensitive to atmospheric contamination until sealed. The equipment offered for perovskite or organic thin-film cell deposition contrasts with the PECVD kit for thin-film silicon as it does not necessarily require gaseous feedstock or vacuum chambers. For example, infinityPV ApS, a Danish company, specialises in flexible polymer solar cells as well as in offering rolls of cells; they also promote small-scale laboratory/prototyping roll-to-roll coating equipment. Options within the basic system allow for slot-die coating, knife coating, gravure and flexographic printing, as well as curing ovens, laminating and testing. These processes work in the open, without low-pressure or vacuum chambers, although an inert atmosphere enclosure may be required for innovative ink formulations, such as

FIGURE 7.2 APPLIED SMARTWEB® roll-to-roll sputter coater for ITO and other conductors, by permission of Applied Materials Inc.

FIGURE 7.3 APPLIED TOPMET™ vacuum evaporation coating system for packaging and flexible electronics, by permission of Applied Materials Inc.

for perovskite cells. A great deal of proprietary knowledge lies behind the formulation of these specialist inks so that they can be laid down and cured without esoteric substrate materials. Another company, Heliatek, uniquely uses vacuum thermal evaporation of successive materials in their R2R production line for organic PV on plastic, similar to manufacturing organic light-emitting diodes.

Some form of hermetic seal will be essential for PV longevity although not all photoactive materials are as sensitive to the ambient as organic PV, and transparent conducting oxides do offer some abrasion resistance without an additional layer. The best-performing encapsulation uses multiple layers of transparent organic and inorganic compounds, so that defects in one layer, such as pinholes, are blocked at their interface by the adjacent layer and do not thread through the entire stack. Dense layers are best deposited by energetic processes, such as plasma-assisted silicon nitride from silane and ammonia gases and plasma polymers from hexamethyl disiloxane (HMDSO) vapour, rather than thermal evaporation. Less dense polymer layers may be deposited by high-speed (tens to hundreds of metres per second) liquid slot-die coating, but this needs a different piece of equipment to vacuum chamber coating. A fast process that is more compatible with vacuum coating is initiated-CVD, in which a deposited monomer is polymerised by a gaseous initiator that is dissociated in the chamber. This may be combined with atomic layer deposition (ALD) of inorganic films (e.g. oxides) for very low rate impervious barriers.

Turning briefly now to the fibre-shaped PV cells whose structure was discussed in Section 5.2.2, it is obvious that few of these research reports have considered any large-scale manufacture of their structures. Contrary to this, a manufacturing scheme was suggested in 2010 by textile engineers from Turkey for fabricating organic PV fibres by R2R (Bedeloglu et al., 2010) in a mostly liquid-based sequence, apart from a metal electrode and antireflection layer, noting the mechanical restrictions in converting this fibre into yarn and then into textile. A recent study of the progress in manufacturing electronic textiles notes that 'the success of fibre-based electronics will depend on the development of feasible and scalable fibre processing and the adoption of industrially viable manufacturing technologies' (Seyedin et al., 2021), whilst also requiring high-speed coating of up to 1000 m/minute. One possible candidate is continuous coaxial fibre drawing, in which the required layers would be relatively thick coaxial coatings on a former that would be drawn down to fibre size.

Another factor influencing the decision on the choice of materials and fabrication technique is the ease with which these devices may be reused at their end of life, enabling materials to be recovered for further uses. Related to this aspect is the amount of energy embedded in PV arrays from their manufacture, which should be recouped during their operating lifetime if they are to be sustainable energy sources. These environmental issues are discussed in the final sections of this chapter.

7.4 ENVIRONMENTAL IMPACT

7.4.1 Energy Payback Time

It is nowadays recognised that an important aspect of any large-scale renewable energy process involves a life cycle analysis (LCA) of it. To set up a renewable energy process – or indeed any other commercial process – requires an input of energy. It is therefore important to estimate how long it will be before the system has returned the overall energy used in its manufacture. For a PV cell, this energy payback time (EPBT) depends on the one hand on its fabrication and structure and on the other hand on the solar climate in which it is being used and its efficiency in converting light into electricity. Other measures have also been proposed, but EPBT is the parameter most widely adopted. The environmental impact of a process is also important to assess, so an LCA must also take account of a number of environmental aspects, as discussed later in this section. In this way, the environmental impacts of different types of solar cells can in principle be compared.

It is evident then that, in order to calculate an EPBT, we need to determine how much energy is used overall in the production of a PV textile fabric. Production includes the formation of the fabric, addition of electrically conducting material, deposition of the PV layers and, normally, addition of an encapsulant. However, as highlighted in Chapter 3, textile fabrics are heterogeneous materials that can be made from a wide range of fibres and for which there are a variety of constructions. Hence, the determination of the total energy required to construct a PV fabric is complex, and any estimate of EPBT is liable to be subject to appreciable error.

Estimates of EPBT were first made 20 years or so ago by various workers for conventional single-crystal and multicrystalline silicon solar cells (Palz & Zibetta, 1991; Alsema, 1998; Knapp & Jester, 2000). The estimates range from 2 to 4 years. Although this represents a wide range of values, these times are still commercially realistic in that the cells are expected to function for about 30 years. Estimates of EPBT were also made for thin-film PV cells (Palz & Zibelta, 1991; Alsema, 1998). In this case, there is much less semiconducting material in each cell, and a greater proportion of the overall energy consumed goes into production of the substrate film and the subsequent PV deposition processes. Estimates for EPBT of 1–3 years were calculated. Again, there is a wide variation in the values determined, but nonetheless the EPBT appears commercially realistic.

More recently, an extensive review (Peng et al., 2013) has collated values for EPBT obtained from a variety of sources. The PV cells considered also included thin-film calcium telluride (CdTe) and CIGS cells. The EPBT values collated were found to vary extensively for any particular PV device, and these variations can be attributed to such factors as manufacturing process, installation methods and conditions at the location of a device. Nevertheless, it can be concluded from this study that the EPBT for CdTe cells is the lowest for all the cells, and the EPBT for amorphous silicon cells appears to be slightly the highest. Overall, however, the values obtained were still less than 3 years, and in a few cases, such as for CdTe, less than 1 year. These studies, therefore, reinforce the case that PV cells are commercially viable in terms of EPBT.

We should also mention organic PVs and perovskites in this context. If the practical difficulties that arise mainly from their susceptibilities to moisture and oxygen can be resolved, both types of PV cells are likely to have significant commercial potential (although the lead content of perovskites may also be a commercial disincentive). Indeed, it has been shown that constructions of both types of cells can possess EPBT values of less than 6 months (Tsang et al., 2015; Gong et al., 2015).

It would clearly be useful if the approaches used to estimate EPBT for thin-film PV devices could be extended to solar textile fabrics. However, there are a considerable number of obstacles. Whilst some solar textile fabrics are produced by a succession of deposition processes onto a fabric (analogous to the construction of thin-film PV devices), some other fabrics are already rendered electrically conductive before semiconductor layers are deposited, and others still are constructed from PV yarns (see Chapter 5). Moreover, in addition to the wide variety of fabric constructions, the origins of the constituent fibres need to be taken into account. In the case of natural fibres, the extraction of the fibres from a plant or animal as well as a variety of sub-sequent preparative stages, such as scouring and bleaching, has to be considered. It has been argued, too, that plant cultivation and animal husbandry should also be included. With synthetic fibres, not only must the production of polymer from the constituent monomers be included, but so too must the preparation or extraction of these monomers.

Whatever their construction, there will be a requirement to estimate values of EPBT for solar textile fabrics and compare these values with analogous solar cells on a glass or polymer film substrate. We have made one attempt for a solar fabric consisting of a standard woven polyester (PET) fabric onto which PV and conduct-ing layers have been deposited (Lind et al., 2015). Despite the paucity of information regarding the overall consumption of energy in producing the woven polyester that was used, it could still be shown that the overall energy consumed to produce our solar fabric was lower than that for glass or thin film. EPBT was estimated to be 1–2 years for a cell efficiency of 5%–10%, and hence comparable to those of more conventional solar cells. Increases in efficiency would lower the EPBT.

7.4.2 GREENHOUSE GAS EMISSIONS

In addition to EPBT, an LCA must also take account of several environmental aspects: these include resources, emissions of greenhouse gases (GHGs) and particu-late matter, waste and destiny at the end of the device's operational life. At first sight, PV technology would appear to have hardly any environmental impact at all. During the operation of a PV device, there are no GHG emissions into the atmosphere, no waste products arising from its operation and no moving parts that require periodic maintenance. However, during its complete life cycle, a PV device does certainly impact on the environment, and PV textiles are no exception. For instance, account has to be taken of the manufacturing processes for the textile, the semiconductor components, the electrically conductive material and any encapsulant. These pro-cesses will produce waste and emit GHGs. So too will the assembly of the solar fabric, its transportation to a desired location, its installation at that location and, at the end of its operational lifetime, recycling or disposal.

A detailed analysis of each of these factors is beyond the scope of this book, but some of them can be usefully discussed here. The emission of GHGs can in prin-ciple be compared on a quantitative basis. The best known GHG is probably carbon dioxide (CO_2), but there are many others such as nitrogen oxides, sulfur dioxide and methane that, weight for weight, have much greater environmental impact. Since the biggest emissions are of carbon dioxide, the greenhouse impact of any other emitted gas is often defined relative to that of carbon dioxide and is conveniently expressed as a CO_2-equivalent amount. GHG emission rates can be assessed per unit of electrical energy generated and so, in principle, be compared for different PV devices.

In their review, Peng et al. (2013) also compared GHG emission rates for various solar cells they were considering. As with EPBT and for much the same reasons, published values for GHG emission rates were found to vary extensively for any given PV device. Moreover, the performances of the devices could not be compared with any certainty, except that the GHG emission rates from CdTe cells did appear lower overall. However, it should also perhaps be noted that there are nowadays some doubts about the use of CdTe in view of concerns in particular over the toxicity of cadmium.

The GHG emission rates published for organic PV devices are also subject to wide variation, but overall it can be concluded that organic PV devices possess low GHG emission rates. Indeed, the low emission is publicised by the German company, Heliatek, for the organic PV film that the company produces (www.heliatek.com/en/technology/sustainability/). The GHG emission rates for perovskites appear significantly higher, but it has been argued that the higher values are to a large extent the result of currently low operational lifetimes, only about 2 years (Gong et al., 2015). An increase in operational lifetime, which perovskites would require to become commercially viable, could lower GHG emission rates considerably.

In turning to the environmental impact of a solar textile, we no longer have to take account of the glass substrate and aluminium mounting needed to support a crystalline silicon cell or the polymer film that supports thin-film cells. Instead, the environmental impact of the textile fabric now has to be considered and indeed this impact can be considerable. We may take as examples cotton and polyester. Amongst natural textiles, cotton is by far the most dominant in commercial terms; amongst synthetic fibre textiles, polyester dominates. In the preparation of woven cotton fabric, several processing procedures are used to remove impurities. So-called 'size', added to warp yarns to prevent fraying during weaving, is removed by hot water with maybe also some mild alkali and detergent. The fabric then has to be scoured with more alkali, usually sodium hydroxide solution, to remove a variety of fats, waxes, seed husks and other impurities. The fabric may also need to be bleached, achieved with hydrogen peroxide. Not only are a variety of chemical agents needed to prepare cotton fabrics but extensive volumes of water are used as well. Several preparative stages are also required for producing wool fabrics and again large amounts of water are needed. In the construction of polyester fabrics, the environmental impacts of synthesising the two constituent monomers (still almost exclusively derived from fossil sources, although synthetic routes from natural sources are now available), reaction between the two monomers to form the polymer, subsequent conversion of the polymer product into yarn, and then conversion of yarn to fabric must all be taken into account.

7.4.3 Decommissioning

It is also worth considering the decommissioning of a solar textile fabric at the end of its operational lifetime. The life expectancy of conventional solar cells is considered to be about 30 years, although it may be a little longer where reduced PV efficiency is acceptable. There is, therefore, the real danger that millions of tonnes of PV waste could have been committed to landfill by 2050. The lifetimes of solar textiles would be expected to be similar, provided that the textile fabric base remains stable over this period. Clearly, more environmentally acceptable alternatives have to be sought, and these alternatives need to become standard practice within the next 10–20 years.

In 2016, an in-depth report on the 'End-of-Life Management of Solar Photovoltaic Panels' was published by the International Renewable Energy Agency (IRENA) and the International Energy Agency Photovoltaic Power Systems. According to this report, the adoption of a circular economy and the well-recognised waste reduction principles of 'reduce, reuse, recycle' can be applied to standard PV cells, and these concepts will have to be extended to PV textiles. The type of textile fabric is determined by the application for which it is eventually destined (discussed in Chapter 8). Scope for the reduction of materials in PV textile fabrics will be influenced by the construction of the fabric. Where the PV and electrically conductive elements are made up of thin layers deposited on a fabric base, reduction of these elements would seem hard to achieve. Where conductive yarns have been integrated into the fabric structure, these yarns have to be thin enough to be successfully accommodated in the fabric, yet thick enough that they do not break when the fabric is bent. In some cases, textile yarns are first rendered conducting before fabric construction (Chapter 4). The use of sufficiently small amounts of metal or conductive polymer would be desirable not only environmentally but also economically, provided good conductivity is achieved. The same type of argument can be given for those cases where the yarns are rendered PV as well as conductive prior to fabric construction (Chapter 5).

Reuse of decommissioned solar panels to provide power elsewhere is gaining some traction. It may be possible for a panel to be repaired and reused in its existing application. If not, there are other options. In some cases, installers seek components of older PV devices to replace faulty ones in existing devices. This approach can provide only a partial solution, in that the fate of the faulty component then has to be decided. In other cases, the entire device may be suitable for another, perhaps less demanding application. For instance, a PV device whose efficiency has reduced over the years may well be suitable for projects to provide power in more remote or less developed locations. PV textile fabrics would be especially suitable, as they would be light, could be rolled up for transportation to the desired location and then unrolled at that location.

If reuse is not an option, then a PV device has to be recycled if it is not to be committed to landfill. The recycling of PV devices is now receiving huge attention. Out of the three principal thin-film technologies, CdTe, CIGS and amorphous silicon, CdTe is arguably the most widely used, and indeed recycling processes have been developed by the US company, First Solar, and the German companies, ANTEC Solar and Loser Chemie. Loser Chemie has also developed a process for recycling CIGS devices. All these processes involve initial crushing or shredding of the PV device and then subsequent chemical treatments. Recycling of solar textiles is still, however, very much in its infancy. Any particular challenges presented by a textile fabric substrate will obviously need to be overcome. One challenge, for example, will arise from the recovery of valuable materials dispersed in the textile waste, bearing in mind the highly heterogeneous nature of textile fabrics (Köhler, Hilty & Bakker, 2011).

REFERENCES

Alsema, E. (1998). Energy requirements of thin-film solar cell modules – A review, *Renew. Sustain. Energy Rev.*, 2, 387–415.

Bedeloglu, A.C., Demir, A., Bozkurt, Y. and Sariciftci, N.S. (2010). A photovoltaic fiber design for smart textiles. *Text. Res. J.*, 80, 1065–1074.

Gao, Z., Bumgardner, C., Song, N., Zhang, Y., Li, J. and Li, X. (2016). Cotton-textile-enabled flexible self-sustaining power packs via roll-to-roll fabrication. *Nat. Commun.*, 7, 11586 (15 pp.).

Gong, J., Darling, S.B. & You, F. (2015). Perovskite photovoltaics: Life-cycle assessment of energy and environmental impacts, *Energy Environ. Sci.*, 8, 1953–1968.

Knapp, K. and Jester, T. (2001). Empirical investigation of the energy payback time for photovoltaic modules, *Sol. Energy* 71, 165–172.

Köhler, A.R., Hilty, L.M. and Bakker, C. (2011). Prospective impacts of electronic textiles on recycling and disposal, *J. Ind. Ecol.*, 15, 496–511.

Li, Q. and Zanelli, A. (2021) A review on fabrication and applications of textile envelope integrated flexible photovoltaic systems. *Renew. Sustain. Energy Rev.*, 139, 110678 (17 pp.).

Lind, A.H.N, Mather, R.R. and Wilson, J.I.B. (2015). Input energy analysis of flexible solar cells on textile, *IET Renew. Power Gener.* doi: 10.1049/iet-rpg 2014.0197 (6 pp.).

Palz, W. and Zibetta, H. (1991). Energy payback time of photovoltaic modules, *Int. J. Sol. Energy* 10, 211–216.

Peng, J., Lu, L. and Yang, H. (2013). Review on life cycle assessment of energy payback and greenhouse gas emission of solar photovoltaic systems, *Renew. Sustain. Energy Rev.*, 19, 255–274.

Schubert, M.B. and Merz, R. (2009). Flexible solar cells and modules. *Philos. Mag.*, 89, 2623–2644.

Seyedin, S., Carey, T., Arbab, A., Eskandarian, L., Bohm, S., Kim, J.M. and Torrisi, F. (2021). Fibre electronics: Towards scaled-up manufacturing of integrated e-textile systems. *Nanoscale* 13, 12818–12847.

Tsang, M.P., Sonnemann, G.W. and Bassani, D.M. (2015). A comparative human health, eco-toxicity, and product environmental assessment on the production of organic and silicon solar cells, *Prog. Photovolt.: Res. Appl.* doi: 10.1002/pip.2704 (11 pp.).

8 Applications of Solar Textile Fabrics

8.1 INTRODUCTION

Over the past 10–20 years, there has been a steadily growing commercial interest in applications for PV textile fabrics. It has yet, however, to take off fully although, as we argue in the next chapter, there is a real likelihood that it will increase sharply soon. One reason that can be posited for the slow, but sure growth in commercial interest is that the application of textiles to generate electrical power has hitherto seemed really quite esoteric and perhaps something confined to being a laboratory curiosity. However, with the increased acceptance of renewable energy and ever-growing concerns about climate change, attitudes are clearly changing and there is much broader support now for new approaches. Some sectors, such as the military for example, have nevertheless been exploring the practical possibilities of solar textiles for a number of years.

8.2 CLOTHING

Much of the interest in solar textiles over the past decade or so has arisen from the growing popularity of wearable electronics. Previously, wearable electronics consisted mainly of medical devices such as hearing aids and pacemakers, but they have also now evolved into fashion accessories like smart watches and activity monitor bracelets. It is perhaps not surprising then that textile items have also been explored and that PV fabric technology has played a significant role in these explorations. Some well-known designer brands have been involved. For instance, more than 10 years or so ago, Tommy Hilfiger (Amsterdam, Netherlands) and Zegna Sport (Milan, Italy) designed solar-powered jackets. A solar jacket is shown in Figure 8.1. Sheila Kennedy in the USA designed PV curtains, using organic PV cells. There is interest too in designing LED-lit garments and drapes powered by thin-film batteries charged from PV cells. Backpacks have been designed by Ralph Lauren (New York City, USA) and Pauline van Dongen, also based in the Netherlands. Over the past decade in fact, Pauline van Dongen has been a champion of solar clothing: she has designed dresses and shirts in which PV cells are incorporated as deliberate design features. Her work very much exemplifies the benefits of designers and technical experts working together on solar textiles. She has received technical support from Holst Centre, Eindhoven University of Technology, the French solar thin-film company Armor PV and Solliance, the Dutch thin-film solar technology research centre. Another example is the collaboration at the University of Wisconsin-Madison, USA, between the designer, Marianne Fairbanks and the chemist, Trisha Andrew (now at the University of Massachusetts Amherst, USA), who have been devising ways of

DOI: 10.1201/9781003147152-8

FIGURE 8.1   The back of a solar jacket. (Reproduced courtesy of Pvilion, New York.)

incorporating dye-based solar cells into various fabrics, including silk and nylon. In a case that has recently come to our attention, the designer and the technical expert are the same person, Kitty Yeung, who has launched her company, Art by Physicist. Her overcoat and dress both feature semi-transparent organic PV films, produced by Armor. We can soon perhaps expect haute couture to embrace the PV effect as a matter of course, given the drive to live a more eco-friendly lifestyle. What enlightened fashion house would not want the fashion scene to charge up the world – literally?

Nevertheless, another key factor impeding faster recognition of the advantages of solar textiles does stem from the clothing and furnishing sectors. For successful marketing, these sectors believe that consumers will naturally want to treat solar textile items in the same way that they would treat traditional textile products. Questions repeatedly asked include: can I wash it? If so, how many times can I wash it? How comfortable is it to wear? In addition, features that are normal in clothing and traditional household textiles, such as drape and stretch for example, would be demanded. Also critical would be the capability of the conducting and PV layers to withstand continuous flexing and creasing, such as at elbows in sleeves and at knees in trousers.

The designs highlighted above utilise in many cases the attachment of PV cells to fabric in some way, rather than the actual integration of the cells in the fabric. As noted in Chapter 4, a particular problem with producing PV fabrics through the integration of PV cells arises from the contrasting properties of textile fabrics and electrically conducting materials: thus, metals possess high conductivity but much lower flexibility and extensibility, whereas conducting polymers are generally flexible and extensible, but their conductivities are 3–4 orders of magnitude lower than those of metals. The approach outlined in Chapter 4 of applying first a layer of conducting polymer and then depositing a metallic layer does go some way towards reconciling these different properties. The incorporation of silver nanowires is also a useful strategy in this respect.

A limitation to the electrical power that PV clothing can generate arises from the limited area of fabric that would be exposed to light. This area would normally be 0.1–0.3 m². Thus, the power generated from an array of solar cells possessing, say, 5% efficiency (a reasonable expectation for amorphous silicon cells, for example) would be 50 Wm⁻², when the cells are exposed to standard solar irradiation of 1 kW/m⁻², the value when the sun is directly overhead. On this basis, the maximum power output would be 5–15 W. Nonetheless, this output is certainly sufficient for powering devices such as mobile phones and tablets. Just as importantly, tiny sensors embedded in the clothing could also be powered. Such sensors, which are small enough not to compromise a fabric's properties to any significant extent, can be used in healthcare and sportswear, for example, to monitor heart rates and respiration rates.

8.3 TENTS AND CANOPIES

Alongside the developments in photovoltaic clothing, a wide range of other applications of solar textile products are likely to blossom in the next few years. Many of these applications feature existing textile fabric products with large areas, which could provide the additional benefit of generating electricity. Notable examples include tents, marquees, awnings and canopies, including canopies for carports. Not only would these products provide shelter and shading but they would also be sources of electrical power. This feature could be very helpful for disaster relief in remote locations, for instance, especially if the transportation of conventional heavy cumbrous equipment would be challenging, such as in an earthquake zone.

There are in fact already a number of examples of solar-powered tents and canopies, and some of these are now discussed. Indeed, several examples first appeared 10–20 years ago. One example of which we became aware about 15 years ago was a solar textile pavilion designed by architect, Nicholas Goldsmith. The skin of the tent consisted of amorphous silicon cells that were encapsulated and laminated to contoured panels on the woven fabric. Since then, this technique has been widely used in producing many other solar fabrics. Amorphous silicon cells are claimed to be particularly effective during the early and late times of the day when sunlight hits the cells at a shallow angle.

A significant player in the solar tent market is the US company, Pvilion. Their tents have attracted particular attention from the US armed forces, who recognised that the tents could be readily erected in remote locations where there was no electrical grid supply. Pvilion also markets fabric solar canopies, including a small solar canopy resembling a sail, as shown in Figure 8.2. These canopies, which utilise polyester fabric coated with PVC, can be installed for outdoor events and in public spaces. Electrical power generated from the canopy can charge mobile phones and laptops in locations where there is no convenient access to a conventional power point.

The US military have also been active themselves in developing solar fabrics. The US Army Natick Soldier Research, Development and Engineering Center in Massachusetts has developed the 'PowerShade' that can provide soldiers protection from the sun and, it is claimed, can generate up to 1.8 kW of electrical power.

FIGURE 8.2 Single-pole solar sail. (Photograph by Michael Gonyea, courtesy of Pvilion, with permission.)

The 'PowerShade' is typically placed over a shelter for billeting soldiers, and applied for charging laptops; and the batteries are used by infantry personnel to power their radios.

Another solar-powered tent is the 'Chill 'n Charge' tent, conceived by the French telecom company, Orange, and the US design company, Kaleidoscope Design, for use by attendees of the Glastonbury Festival, the popular festival of performing arts held annually in southern England. The tent possesses three directional glides that can be adjusted throughout the day to maximise the collection of solar energy. Photovoltaic threads are interwoven with a standard test fabric. It is claimed that the tent can power many devices simultaneously, including an under-floor heating element, triggered when the temperature inside falls below a pre-set level. At the heart of the tent is a control hub that can emit a wireless internet signal. In addition, lost campers can find their tent by sending a text message, which activates a glow in the tent that helps them to locate it.

Other solar-powered tents on the market include the Eureka 3-Season Tent, containing flexible solar panels positioned on top of the tent, and the Big Bang Solar tent, in which a solar panel that is slid into a pouch on the tent's exterior charges a bank of lithium batteries. The KATABATIC Solar Power Tent and the High-Quality Solar-Powered Tent are examples of solar-powered tents designed for durability and rugged weather, especially for expeditions. It is evident therefore that solar-powered tents are now becoming quite standard items, and their appeal, we believe, is likely to grow still more in the future.

Retractable solar canopies have also been developed. An early example is one designed by the German company, Warema Renkhoff, nearly 20 years ago. Energy is collected from the solar cells and stored. The stored energy is used to operate the winding drive motor.

A later development came from Sheila Kennedy and her collaborators. For the past 10 years or so, she has expounded the concept of 'Soft Architecture'. Two key aspects of her concept are design that enables interaction between materials and digital networks and an infrastructure that can adapt to changing conditions. This infrastructure can use a variety of energy sources, one of which is PV energy. Her concept is illustrated, for example, by the addition of thin-film organic solar cells to textile-based roof canopies, situated in dense urban areas, as demonstrated by a case study in a district of Porto in Portugal. Each house possesses an interior stairwell, providing daylight and ventilation. A robust rooftop photovoltaic textile canopy was designed, such that it could be combined with the vertical space of the stairwell. During the day, the canopy can provide energy and shading. At night, the canopy is retracted and rolled into the stairwell. The energy generated during the day is stored in the stairwell and then used to power lighting inside and outside each house.

8.4 TARPAULINS

Related to the development of solar tents and canopies is the development of solar tarpaulins. Although tarpaulins are used particularly as covers for goods transported by lorries and semi-trailers, they also have many other practical uses: e.g. providing temporary shelter and roof repairs in disaster zones; protecting building materials on construction sites; using groundsheets and windbreaks for camping. The area of tarpaulin in a fully laden semi-trailer could be up to 50–100 m², and so several kW of electrical power could, in principle, be generated during daylight hours, even when it is taken into account that at any particular time different sections of the tarpaulin would be exposed to different intensities of light. Clearly, the PV power generated would be insufficient for driving the vehicle, but, according to Jonas Sundqvist at the Fraunhofer Institute for Ceramic Technologies and Systems (IKTS), it can certainly contribute to the supply of heating and cooling units (Sundqvist, 2019). IKTS envisages thin film PV films being attached to standard lorry tarpaulins, constructed from a glass fibre fabric. Tarpaulins for goods vehicles are generally constructed from polyester fabrics coated with polyvinyl chloride (PVC). A step forward will be the integration of solar cells on such fabrics, and indeed on any successor fabrics deemed more environmentally sustainable.

Other examples can also be cited. The French company, Armor, produces an organic PV film, ASCA, that can be incorporated into retractable covers for electric vehicles. The start-up company, Solar Cloth, also based in France, has developed a PV tarpaulin for lorries built by Renault. Tarpon Solar AS, in Norway, has developed a type of tarpaulin on which are laminated solar cells produced by the Swedish Company, Midsummer AB.

8.5 ARCHITECTURE

The need to render buildings much more energy efficient is now widely recognised. One strategy towards achieving this goal is the adoption of building-integrated photovoltaic (BIPV) technology, which can even transform a building from being an energy consumer to being a net energy producer. Conventional glass solar panels are now of course a common sight on the roofs of domestic houses and other buildings, but, as explained in Chapter 1, they suffer from being fragile, rigid and heavy, and can only be laid on flat surfaces. These drawbacks have been partly overcome by the application of thin solar films. For example, the German company, Heliatek, produces organic PV films on an R2R production line, with successive layers deposited using thermal evaporation processes. 500 m^2 of their film was installed on the roof of the Pierre Mendès France school at La Rochelle on the west coast of France. A recently launched British company, Solivus, installs Heliatek's films on commercial buildings and other constructions with large roof areas, and the organic PV film, ASCA, produced by Armor, can be similarly applied.

Nevertheless, these solar films do have some disadvantages that solar textile fabrics would overcome. The thin nature of the films can render them susceptible to fracture, during both processing and subsequent use, whereas woven fabrics in particular do not have this problem. In addition, whilst thin films clearly possess the flexibility that is absent from conventional solar panels, a textile fabric can be constructed to provide whatever flexibility is most suitable for a particular roof. Indeed, a fabric could be tailor made with different flexibilities in different parts of the fabric for installation on a roof with a complex surface geometry.

Textile architecture is becoming more prevalent too in permanent- and semipermanent-tensioned membrane constructions, such as exhibition halls, concert venues and sports complexes. A tensioned-membrane construction is a fabric structure supported by beams and masts. Examples include the tensioned fabric membrane structures of the roofs of Dynamic Earth in Edinburgh (see Figure 8.3) and the O2 Arena in London in the UK, and also at Denver International Airport in the USA. These roof structures are gaining popularity because they are lightweight and translucent, in contrast to traditional roofs, and they possess a sleek, aesthetic appeal. Such fabrics cover large areas and, if made PV, would be sources of considerable power. Dynamic Earth, for example, covers about 15,000 m^2, so several hundred kW of electrical power could in principle be generated.

FIGURE 8.3 Dynamic earth, Edinburgh, showing its tensioned membrane roof structure. (Source: Shutterstock.)

Another feature of a permanent building that lends itself to textile fabric is the vertical outer cladding, which provides thermal insulation and weather resistance. Cladding can also improve the appearance of buildings. Cladding materials include wood, stone, brick and concrete – and increasingly nowadays, textile fabrics, in view of their smaller weight, ability to be rendered flame retardant and lower insulation costs. Textile cladding also provides wide scope for artistic expression. In addition, the fabric can be PV.

Cladding has of course in recent years been very much in the news, especially in the UK, as a result of its role in the fatal fire at Grenfell Tower, a residential tower block in London. Regulations governing the properties and performance of cladding have consequently been tightened considerably, so solar fabrics used in cladding would also have to conform to these tighter regulations. Outstanding flame retardancy has to be a key property.

Vertical façades on buildings tend to be much more subject to shade, especially in crowded conurbations, and so the power-generated per unit area of cladding will generally be less than on roofs. On the other hand, vertical façades cover much more area on a building than roofs do, so there is still an incentive to render façades PV. Growing interest has been reported in the application of organic solar cells and perovskite solar cells to façades, as these types of cells can function efficiently under conditions of low irradiation. The commercial interest in PV textile cladding is bound to increase, especially with its attendant opportunities for creative design.

An unusual concept for PV cladding was designed by the American venture, Kennedy and Violich Architecture, directed by Sheila Kennedy and J. Frano Violich. Their project, called IPA Soft House, includes an innovative use of PV textile fabrics, trialled in a block consisting of four attached houses in Hamburg in Germany. The building, designed on an environmentally aware basis, is constructed of wood and clad in several materials, including an insulating layer of wool. The south side also includes highly visible PV cladding consisting of textile fabric ribbons, to provide shading and to generate electricity, used mostly to power lighting. The flexibility of the ribbons enables them to be twisted, to allow the fabric to track the sun during the course of each day.

Detailed information on the scope for BIPV textiles can be found in a recently published wide-ranging review by Li and Zanelli (2021). These authors suggest that the capability to mass produce suitable PV textile fabrics using R2R processing will determine the level of commercialisation achieved. They also predict that retrofits to existing buildings will be the dominant application for PV textile fabrics.

8.6 AGRICULTURE

The benefits of solar cells are now being increasingly recognised in agriculture and horticulture. They are finding applications in greenhouses, farm buildings and even on the land itself. The way in which these benefits can best be exploited is to quite a large extent influenced by climatic conditions. Let us first consider the use of solar cells on the land. A major problem with using solar panels to assist crop growth has been that they occupy valuable land that would otherwise itself be used for crop growth. A concept known as agrophotovoltaics (APV) has accordingly taken off, whereby the crops are grown underneath mounted solar panels, thus releasing land that would have been occupied by the panels. Clearly, the panels have to transmit sufficient light to the crops beneath them. Beck et al. (2012) and Barron-Gafford et al. (2019) have reported that the benefits of the APV strategy can include higher yields for many (though not all) crops, a reduction in the amount of water needed to irrigate the crops – and even more efficient generation of electricity. A recent European report has noted that APV is well suited for the cultivation of berries, fruit orchards and vines (Solar Power Europe, 2021), and Barron-Gafford et al. (2019) have reported benefits to the cultivation of tomatoes and jalapenos in Arizona, USA. The APV strategy prevents exposure of crops to direct sunlight and hence reduces loss of water from the plants due to transpiration. These aspects are particularly important for crop growth in warmer climates, such as tropical and subtropical regions. The higher PV efficiency has been attributed to the emission of water vapour from the plants, which helps to cool the panels.

Weselek et al. (2019), however, have cautioned that there could by contrast be a reduction in crop yield due to reduced solar irradiation under the panels, although reduced yields under the panels may be counterbalanced by the extra land made available from the use of APV. The intensity of the incident sunlight, therefore, is an important element. Crops may benefit in warmer climates from the shade provided by the solar cells but may not flourish as well in more temperate areas. The extent to which these differences are significant will depend on the type of crop. Thus, the nature of the crop and the climate in which it is grown will be important factors in this respect.

The concept of APV is prompting further studies aimed at optimising the balance between crop production and the PV generation of energy. It will be surprising if robust, flexible materials of lighter weight do not make an impact on APV for many of the reasons already cited throughout this book. In particular, the whole assembly that supports the panels could be more easily raised, if that is required, to allow harvesting of the crop, and also be readily relocated to other areas of land, depending on the nature of the crops and the time of year. PV films would appear to be an attractive option, and certainly such films can be produced that transmit a substantial fraction of the light that falls on them. However, as already highlighted, PV films are generally more fragile and less durable than glass. PV textile fabrics, of the semitransparent type used in tensioned roof structures, would circumvent these drawbacks, especially in locations with more intense sunlight. Alternatively, there is scope for attaching PV films to semitransparent textile fabrics, although care would have to be taken not to compromise flexibility to any significant extent.

The APV concept has also been applied to the cultivation of crops in greenhouses, notably green vegetables, herbs and berries. Cultivation of flowers can also benefit from APV. Greenhouses that include solar panels in their structure can generate electricity to control factors such as humidity and ventilation. Where greenhouses are situated in cooler climates, interest is developing in extending the growing season by a few weeks through the introduction of low-level heating. In warmer climates, the solar panels would again have the advantage of providing shade. As with APV assemblies erected on the land, flexible materials, including semitransparent textile fabrics, are attractive options for supporting the PV cells. They would seem particularly attractive for polytunnels, although it has been noted that the PV modules can be quite severely degraded by mechanical stresses on them, caused by movement of the plastic sheeting, particularly in very windy conditions (Magadley et al., 2021).

In considering the use of photovoltaic generation in agriculture, farm buildings should not be overlooked. For example, they provide storage for equipment, tools and materials; protect livestock during harsh weather; and even provide protection for fish farming. The advantages of solar textiles, already discussed above for buildings in general, will also apply to farm buildings.

A novel application of APV is grazing, notably sheep grazing. Whilst it has been introduced in the USA and many European countries over the past decade or so, it appears now to be taking particular hold in the eastern states of Australia (Australian Guide to Agrisolar, 2021). Extensive work on the development of APV for grazing has been undertaken by the University of Queensland at its solar farm in Gatton, west of Brisbane. As well as generating electricity, solar panels protect the sheep from rain and intense sunlight. The extra advantages that solar textile fabrics could offer are similar to those identified for APV in land crops.

8.7   SAILING

The inclusion of solar panels in sails is now gaining some considerable traction, yet we know of a patent for solar sails issued in 2001 (Muller, 2001). A type of sail was designed in which flexible solar cells are attached to the sailcloth, and it was stated that the cells possessed a flexibility matching that of the sailcloth. The past decade, however, has witnessed much more widespread commercial interest. A major player has been Alain Janet of Solar Cloth in France, who has collaborated with UK Sailmakers (France), whose parent company is based in the USA – despite its name! Layers of CIGS cells as thin as 65 μm are attached to a boat's mainsail, and these layers are encapsulated in order to protect them from harsh environments (see Chapter 7). The encapsulation process is the same as that used for producing the sail fabric. Janet claims that as much as 1 kW of power can be generated, provided cells are installed on both sides of the mainsail. The solar sails were trialled on the boat, Open 50, in the Route de Rhum transatlantic race in 2014. The solar sails were then used on the Arcona 380Z sailing boat, constructed in Sweden, and the American J/88. Aboard both these vessels were large batteries produced by the Finnish company, Oceanvolt, and these batteries were fed directly from the cells on the sails. The batteries could power electric motors, where required, to propel the vessel. Figure 8.4 shows an image of the sailing boat, Spirit 44CR(e), showing the solar cells on the sailcloth.

It is expected that solar power will become increasingly attractive to yachtsmen and yachtswomen and that its appeal will extend to the luxury market. For example, the Norwegian designer, Kurt Strand, has designed a concept for a 160-metre-long solar superyacht that he calls 'Florida', which offers a swimming pool, a helipad and a bar that is two storeys high! However, in those vessels with a motor on board to propel them, a sail may not always be necessary. Thus, solar-powered 'catamarans' have been developed in a collaboration between Solar Cloth and Silent Yachts, whose headquarters are in Austria. Solar radiation is captured by solar panels located on

FIGURE 8.4   Solar sail on the sailing boat, Spirit 44CR(e). (Photograph by Richard Langdon, courtesy of Solar Cloth, France.)

the bridge of the vessel. The power generated, it is claimed, is not only sufficient to propel the vessel but supply onboard appliances as well. Solar fabrics can be useful in this respect, in that they can be rolled up or laid out on board, as required.

8.8   AIRSHIPS

An airship is a craft that is lighter than air yet, like an aeroplane, can be propelled in a desired direction. The gas used for lifting an airship is generally helium. Hydrogen has also been used, but its high flammability has been the cause of a number of airship disasters. There are three main types of airship: nonrigids (often known as blimps), semirigids and rigids. Figure 8.5 provides some examples. All contain balloons, usually made of polyester or maybe neoprene. A blimp is essentially a massive balloon that contains the gas, with a gondola slung underneath for carrying passengers and crew. A semirigid is very similar, but it possesses a structural keel that extends along the length of the base of the balloon and supports the gondola. A rigid airship possesses a framework of metal girders covered with fabric. It is not, however, airtight; inside the framework are several balloons filled with gas. Perhaps the most remembered examples of rigid airships were those designed by von Zeppelin in Germany in the early twentieth century and the Hindenberg, which dramatically exploded in 1937 while attempting to dock at Lakehurst Naval Air Station in New Jersey in the USA.

(a)   (b)

FIGURE 8.5   Airships: (a) non-rigid (blimp); (b) rigid. (Source: Shutterstock.)

The appeal of airships then waned considerably, although the US Navy used blimps for anti-submarine operations in the Second World War, and since then blimps have been used quite widely for advertising. The popularity of airships is now showing signs of a resurgence, however. For example, the British company, Hybrid Air Vehicles, has been developing an airship for short-distance intercity travel, to be in operation by 2025. It is envisaged too that airships will be used extensively for transporting cargo. For example, several years ago, the American company, Aeros, announced the start of production of a nonrigid airship for transporting cargo, and the French company, Flying Whales, is developing an airship that can lift 60 tonnes. One problem with these developments may arise from shortages of helium. Hydrogen, although lighter and more abundant than helium, is still considered too risky to use.

The power required to drive an airship is only 10%–20% that needed for conventional aircraft. There is consequently a clear attraction in harnessing solar power to propel an airship. Indeed in 2009, a team of French students designed a solar-powered blimp, Nephelios, with a flexible array of solar cells attached to the balloon. It was planned to fly Nephelios across the English Channel exactly 100 years after Louis Blériot's pioneering flight. Although this plan was never realised, Nephelios did nevertheless succeed in flying on solar power alone.

Currently, Varialift Airships in the UK are developing a solar-powered airship at a commercial level. The airship possesses an aluminium frame and is to be powered directly by two solar-powered engines and two conventional jet engines. No onboard battery is envisaged. Other companies developing similar craft include Lockheed Martin in the USA and Flying Whales. The worldwide emphasis on the elimination of carbon-based fuels may well see solar airships being used quite extensively on a commercial basis within the next 10 years. There is clearly then an opportunity for the implementation of solar fabrics, in that they can be specially tailored to fit onto the curved surfaces of airships.

Stratospheric airships are also attracting growing interest. Despite the development of satellites for ever more sophisticated communication systems, a high cost is involved in the launch and deployment of a satellite. Furthermore, the satellite population is now very crowded, and the extent of debris in their path puts them at risk. On the other hand, ground-based communication systems have limited range and can be prone to distortion and even failure. As a consequence of these drawbacks, airships operating in the stratosphere at heights about 20 km above the earth's surface are attracting increasing attention, both in civilian and military areas. Blimps appear to be the most favoured, because of their lightness compared to other structures.

A drawback of operating in the stratosphere arises from its chemical activity, particularly from the presence of ozone, which acts to restrict the intensity of ultra-violet radiation reaching the earth's surface from the sun. However, it is also recognised that the intense solar radiation provides an opportunity for generating electrical power through solar panels attached to the exterior of an airship. The energy produced can then be used to propel the airship during the day and also be stored, so that the airship can be driven at night. Nevertheless, in addition to the resistance to ultraviolet radiation, resistance to attack by ozone is a key consideration.

8.9   AESTHETIC ASPECTS

In every application that has been discussed in this chapter, the solar textile will necessarily be very visible because it has to be placed on surfaces available to incoming light. Indeed, this is true of any solar panel, and the glass panels that are fitted onto rooftops are very obvious examples. Yet, to many eyes, whilst the benefits of such panels are evident, the panels are really quite ugly. The architect, Nico Rensch, who is credited with designing the UK's first energy-positive factory-built house, is quoted as commenting (York, 2021):

'I hate solar panels. They are a crime in terms of aesthetics'.

There is therefore a growing desire – and, arguably, even a need – to make solar panels more aesthetically pleasing, and the use of textiles in photovoltaic technology is beginning to meet this need. For example, arrays of textile-based solar cells can be designed to blend in with a roof or, alternatively, provide a special visual feature.

Already, a quest to mix aesthetic qualities and technical performance in solar panels generally is gaining some momentum. Art and Energy, a company based in Exeter, UK, designs solar panels as art materials and explores the artistic opportunities inherent in PV technology. In the Holy Family Cathedral at Saskatoon in Canada, there are stained glass windows in which are embedded more than 1000 solar cells. Given that textile fabrics have for centuries, and even millenia, been judged on their aesthetic appearance, there is clearly scope for solar textiles to enhance the aesthetic appearance of solar panels.

The development of aesthetic aspects of solar textiles is still very much in its infancy, something that can, at least partially, be attributed to the problem of compromising the very different physical properties of solar cells and textile fabrics – as discussed in Chapter 4. Perhaps unsurprisingly, the most manifest developments have been in the clothing sector, and we have already referred to the progressive impact of solar textiles on this sector. Quite recently, design and technology researchers at Aalto University in Finland have been collaborating on a 'Sun-powered Textiles' project, with a view to creating original clothing with concealed solar panels. The panels power small sensors, such as temperature and humidity sensors present in a garment. The concealment of the panels is to ensure that the appearance of the garment is compromised as little as possible, thus allowing greater freedom of design.

This approach, at first sight, seems absurd, in that a solar cell obviously requires light to activate it. It is stated, however, that the fabric is designed to permit enough light to impinge on the panels for them to power the sensors. To achieve sufficient penetration of light requires modifying the physical properties of the textile and any finishing treatments applied to it. Applications are envisaged to include not just fashion but also workwear and sportswear and even curtains and screens.

An interesting project is being undertaken by Kathleen McDermott and her colleagues at New York University in the USA (McDermott et al., 2021). They are seeking to widen educational content about solar panels to include not just technological aspects but also aesthetic aspects. They document various methods of embedding solar panels into a fabric with a view to studying the different aesthetic outcomes. They demonstrate these different outcomes with solar panels embedded in a scarf; the panels power LED messages. These messages, the authors state, are intended to help in overcoming problems of communication that can arise with the wearing of masks and face coverings.

In this context, we may perhaps also allude to our own work, in which amorphous silicon cells are integrated into woven polyester fabrics (Diyaf et al., 2014). An example of our solar fabric is shown in Figure 8.6

Some developments in the aesthetics of solar textiles in buildings should also be noted. We have already referred to the innovative designs in buildings by Sheila Kennedy and her colleagues and to the possibilities offered by tensile membrane fabric roof structures. Arguably, the Allianz Arena football stadium in Munich in Germany should also be mentioned (Figure 8.7): it has a huge fluoropolymer textile membrane structure that renders the stadium fire-resistant, and some solar cells are embedded in the structure.

Developments in the aesthetics of solar textiles in other sectors still appear largely embryonic at the moment. Nevertheless, we expect commercial emphasis on aesthetics in solar textile applications generally will steadily rise. Indeed, it can be well argued that some of the solar textiles already described in this chapter possess desirable aesthetic qualities.

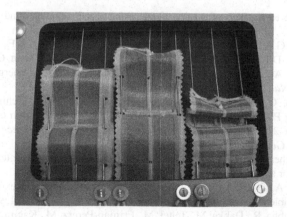

FIGURE 8.6 Woven polyester fabrics with integrated amorphous silicon solar cells, showing the aesthetic potential of solar textile fabrics.

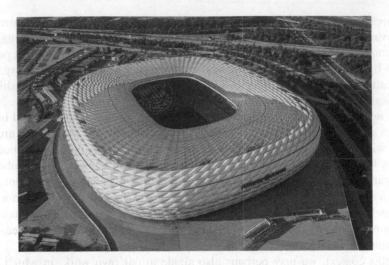

FIGURE 8.7 Allianz Arena, Munich. (Source: Shutterstock.)

8.10 SOME CONCLUDING REMARKS

In this chapter, we have sought to demonstrate the multitude of applications for solar textiles. We do not pretend that the number of applications we discuss is exhaustive, but we hope we have shown just how wide-ranging the applications already are. Some examples, such as clothing items, tents and fabrics for APV, have been steadily gaining traction for a number of years. Others, such as sails and airships, have yet to make such a big impact but, in our opinion, are very likely to do so within the next few years. The aesthetic appearance of solar textiles will also come to be an important factor.

REFERENCES

Australian Guide to Agrisolar for Large-Scale Solar, for Proponents and Farmers (March 2021).

Barron-Gafford, G.A., Pavao-Zuckerman, M.A., Minor, R.L., Sutter, L.F., Barnett-Moreno, I., Blackett, D.T., Thompson, M., Dimond, K., Gerlak, A.K., Nabhan, G.P. & Macknick, J.E. (2019). Agrivoltaics provide mutual benefits across the food-energy-water nexus in Drylands, *Nat. Sustain.*, 2, 848–855.

Beck, M., Bopp, G., Goetzberger, A., Obergfell, T., Reise, C. & Schindele, S. (2012). Combining PV and food crops to agrophotovoltaic – Optimisation of orientation and harvest. In *27th European Photovoltaic Solar Energy Conference and Exhibition, Frankfurt, Germany*, pp. 4096–4100.

Diyaf, A.G., Mather, R.R. & Wilson, J.I.B. (2014). Contacts on polyester textile as a flexible substrate for solar cells, *IET Renew. Power Gener.*, 8, 444–450.

Li, Q. & Zanelli, A. (2021). A review on fabrication and applications of textile envelope integrated flexible photovoltaic systems, *Renew. Sustain. Energy Rev.*, 139, 110678 (17 pp.).

Magadley, E., Kabha, R., Dakka, M., Teitel, M., Friman-Peretz, M., Kacira, M., Waller, R. & Yehia, I. (2022). Organic photovoltaic modules integrated inside and outside a polytunnel roof, *Renew. Energy* 182, 163–171.

McDermott, K., Byun, B., Tiwari, A. & Hu, A. (2021). Solar scarf: Expanding DIY educational content with an expressive wearable system. In *ISWC '21: 2021 International Symposium on Wearable Computers, Virtual Global*, 194–198.

Muller, H.-F. (2001). *Sailcloth Arrangement for Sails of Water-Going Vessels*, US Patent Office, 6237521, May 2001.

Solar Power Europe (2021). Agrisolar Best Practices Guidelines Version 1.0.

Sundqvist, J. (2019). The truck as a solar power plant? Current research at Fraunhofer. https://logistik-aktuell.com/2019/11/05/truck-solar-tarpaulins-research/.

Weselek, A., Ehmann, A., Zikeli, S., Lewandowski, I., Schindele, S. & Högy, P. (2019). Agrophotovoltaic systems: Applications, challenges and opportunities. A review, *Agron. Sustain. Develop.*, 39, 35 (20 pp.).

York, M. (2021). House of the week: Britain's first off-the-shelf eco-house that can actually earn you money, *Times*, October 29.

9 The Outlook for Solar Textiles

9.1 INTRODUCTION

Conventional rigid PV arrays are not in themselves sufficient except in the simplest of applications but require some 'balance of system' components to manage the output power (e.g. an inverter to convert DC to AC and for grid connection) and often to store any energy that is not immediately fed to a load. As we have seen in the preceding chapters, there is a wealth of innovative design and supporting technology for implementing textile photovoltaics, along with a range of applications awaiting commercial realisation. Some of these expect the integration of flexible devices with a PV power supply or with the addition of other complementary energy sources as well as with flexible electrical storage. We have seen that it is already challenging to make textile-compatible PVs, before considering them in combination with other flexible devices! This final chapter considers these additions and their potential uses, although demanding a commercial production route and reduced costs in order to be marketable.

9.2 ELECTRICAL STORAGE

Usually, solar cells are not able to store the energy they generate and so are an intermittent power source, which poses difficulties for many applications unless some form of energy storage can be incorporated: recall Dean Swift's sunbeam-storing cucumbers in Chapter 2! A recent report of a novel photoelectric device that generates electrical charges when illuminated with blue light and stores them in a thin surface layer (Jiang et al., 2021) offers a function that is not presented by conventional photovoltaic cells, but as operation is only at 30 K, this is not yet a viable option for our textile devices. However, dye-sensitized solar cells depend on electrochemical reactions and are not pure photovoltaic devices, as we saw in Chapter 2, thus they are a possibility for fabrication with integral electrochemical storage.

Presently, electricity may be stored in a battery, which operates by storing charge in a reversible electrochemical reaction, or in a capacitor, which does not require any chemical activity and has a longer life. They are used in different situations, with batteries providing higher *energy* density than capacitors, which are generally able to be charged and discharged much more rapidly and have higher *power* density. (For example, wireless communication generally requires some milliamps of current during transmissions, which can be supplied readily by batteries.) Another significant characteristic of capacitors used for energy storage and supply is that their voltage will fall as the charge is extracted, in contrast to batteries which tend to have a constant voltage until nearly discharged, depending on their chemistry (Figure 9.1).

DOI: 10.1201/9781003147152-9

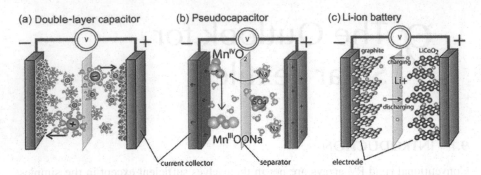

FIGURE 9.1 Schematic diagrams of two forms of supercapacitor and a Li-ion battery, indicating the movement of charges. In (a), the charges are stored electrostatically in a double layer between each of the electrodes and mobile counter ions; in (b), there is the same double layer of charge, together with additional charge storage from reversible chemical surface reactions similar to electrochemical batteries, such as shown in (c) for one type of Li-ion battery. (From Jost, K., Dion, G. and Gogotsi, Y., Textile energy storage in perspective. *J Mater Chem A*, 2014, 2, 10776-87.)

The energy in a conventional capacitor is stored in the electric field produced between metal contacts, when an applied voltage separates pairs of positive and negative charges: the greater the charge, the higher the field. This potential energy is released when the charges are extracted. The charge that may be stored not only depends on applied voltage but also on the capacitance, which is a function of the area of the capacitor contacts and their separation: high capacitance requires large area contacts with a narrow spacing, perhaps achieved in a small package by rolling up a thin sandwich of metals and spacer. Capacitance is also increased by filling the air gap with a dielectric. High-value capacitors were typically filled with an electrolytic paste that would eventually fail as the electrolyte slowly evaporated, but modern solid-state capacitors use ceramic dielectrics with high permittivity.

Supercapacitors go beyond this and in their plainest form contain an ion-permeable separator within an electrolyte, held between porous, activated carbon contacts. When a voltage is applied to the contacts, a double-charge layer forms on each contact by ions from the electrolyte, in effect providing two very narrow capacitors in series: the charge in the surface layer of electrolyte is ionic, separated from the contacts by only a molecular layer of solvent held there by electrostatics. Thus, high capacitance is achieved by a combination of large-area contacts, a very narrow gap and a high dielectric permittivity. The main drawback is that the applied voltage is much more limited than with conventional capacitors as the electric field lies across the ultrathin molecular layer of solvent at each contact and can cause breakdown if increased too far. More complex forms of supercapacitor incorporate electrochemical charge storage in addition to electrostatic charge storage, using asymmetric electrodes.

A comprehensive review in 2016 of textile energy storage covered both one-dimensional yarn and two-dimensional fabric structures for batteries and supercapacitors, many from Chinese researchers (Zhai et al., 2016). The range of materials and constructions shows that there was yet no winning design, partly because of the varied size of demand from the diverse applications. Another requirement for some uses is wearability, which necessitates meeting the challenges of textile industry standards and perhaps washability.

9.3 FLEXIBLE ELECTRICAL STORAGE WITH FLEXIBLE PV

Reviews of flexible PV integrated with flexible electrical storage tend to concentrate on small-scale wearable electronics as the most accessible products with present-day technology (e.g. Varma et al., 2018; Zhao et al., 2021; Devadiga et al., 2022). These potential uses require only low energy but may operate for long periods, perhaps combining a sensor with telemetry as we shall see later in this chapter. Any type of flexible PV is a possible partner for rechargeable batteries (based on either zinc or lithium chemistries) or for supercapacitors (that use advanced materials such as graphene or esoteric metal oxides), but the combination must be durable and meet environmental and safety principles. Several research examples demonstrated stability when enduring bending trials, as well as a reasonable number of charge/discharge battery cycles. Very few of the early examples were fabricated with textile fabrics and instead used polymer sheet or even metal foil substrates.

An account from Berkeley, USA, in Ostfeld et al. (2016) combined thin-film amorphous silicon solar cells on polyimide with a lithium-ion battery having stainless steel and nickel foil current collectors, contained within a laminated polyethylene/aluminium/polypropylene pouch, to power a pulse oximeter. This device monitored blood pressure and oxygen content with LED/photodiode probing, requiring ~20 mA during its measurement phase and demonstrated a successful battery design, with effective solar charging when the load current and photovoltaic current were matched. Another account in the same year also used commercially available amorphous silicon on polyimide solar arrays but combined them with flexible supercapacitors on activated cotton textile by R2R lamination (Gao et al., 2016). The supercapacitor electrodes were 3D-nanostructured metallic-layered double hydroxide positive electrodes and graphene negative electrodes, each on activated cotton, which gave large effective area, and the space was filled with a PVA-KOH gel electrolyte. This asymmetric supercapacitor gave promising charge/discharge performance and worked well even when folded. A LED was powered by the PV when this was illuminated and by the supercapacitor for several minutes when in the dark, demonstrating the effectiveness of the combination. One-dimensional devices based on fibres were reviewed around the same time, concentrating on DSSC with Li-ion batteries or supercapacitors, and noting the long conduction path for this configuration (Sun et al., 2017).

Since that period, interest in integrated PV generation and storage has intensified, as shown by the reviews cited at the start of this section. There is a useful summary in the 2021 review by Zhao et al. of the energy densities of flexible storage devices, ranging from tens of milliwatt h cm^{-2} for Li-ion or Zn-ion batteries to tens of microwatt h cm^{-2} for supercapacitors. As noted above, a DSSC may be integrated with a Li-ion battery by sharing a common electrode. Despite all this activity and sensors for many conditions, there is still a dearth of applications on textiles.

9.4 FLEXIBLE ENERGY HARVESTING COMBINATIONS

The obvious limitation to relying on PV electricity, namely the lack of output in the dark, may be alleviated by combining PV with complementary energy-harvesting mechanisms, perhaps still incorporating some electrical storage. Non-contact wireless charging using some form of flexible conducting induction coil or antenna does work in some situations but has limited applicability. Alternative resources include thermoelectricity, piezoelectricity and triboelectricity. Focussing only on wearables, it is the last two of these that are the most practical, as the thermoelectric effect requires a higher temperature gradient than is comfortable (although a 1% efficient thermoelectric e-textile generated 700 µW/ m^2 from a 7K temperature difference). The other two effects utilise mechanical forces, either deformation by compression or tension (piezoelectric effect), or friction (triboelectric effect). The former is the source of sparks for gas igniters, and the latter is the source of electrostatic charging in some insulating materials. They generally provide a high voltage but very small current, with output powers up to ~2 mW/cm^2 for textile-based triboelectric nanogenerators (TENG). A TENG artificial skin of flexible polymers and metal that may be integrated with PV and storage has lower output than non-textile devices (Komolafe et al., 2021).

A good example of what may be engineered is a combination of fibre TENG made from PTFE strips and fibre DSSC on polymer wires, each being woven into a textile and contacted with copper-coated electrodes in either series or parallel sequences (Figure 9.2). Alternative weave patterns were examined, mixed with wool fibres, to optimise the energy produced and noted that the impedance mismatch between the two types of generators required at least an interconnecting diode (Chen et al., 2016). A few square centimetres of this combination successfully powered a watch or mobile phone and could charge a 2 mF capacitor to 2 V in 1 minute whilst in sunlight and squeezed by handshake. Durability was shown by successful operation during 500 bending cycles although the TENG output was affected by humidity. Another example of integrated fibre DSSC and TENG generators also included supercapacitor storage (Wen et al., 2016). A woven textile comprised a top woven fabric DSSC layer (using Ti wire base fibres) and a bottom woven supercapacitor layer (using carbon fibres), with a TENG formed between these layers by pairs of copper-coated EVA tubing and PDMS-covered copper-coated EVA tubing. A diode is needed to prevent current from the TENG flowing through the DSSC, and a rectifying circuit is needed to convert the alternating current from the TENG into DC to charge the supercapacitor. A small patch of this textile comprising several DSSCs and supercapacitors was attached to a T-shirt together with switches to use either the DSSC output or the supercapacitor charged by triboelectricity without incorporating impedance matching. However, as Dong and Wang note, stacking a TENG on a supercapacitor using the same fabric may have problems with the mechanical force on the TENG affecting the supercapacitor below, therefore requiring some consideration of different structural approaches. Nonetheless, a combination of DSSC and TENG with a supercapacitor in a bracelet can power LED lights, a watch and temperature sensors (Dong & Wang, 2021). The recent review by Devadiga et al. (2022) cited in Section 9.3 above shows several more recent examples of wearable DSSC combined with TENG and either batteries or supercapacitors, concentrating on developments in fibre DSSCs (Figure 9.3).

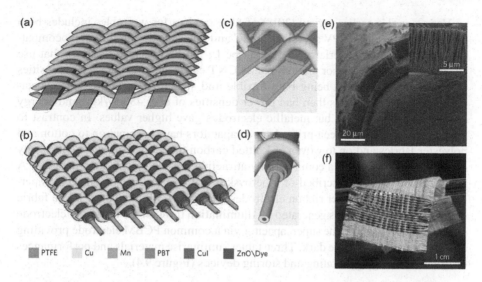

FIGURE 9.2 Power textile combining (a) TENG and (b) DSS-energy harvesting in (f) a woven textile that includes woollen yarn. (c) and (d) show the electrode interlacing for each of these generators using copper-coated polymer fibres. (e) is an SEM of the fine-structured photoanode for the DSSC. (From Chen, J., Huang, Y., Zhang N., Zou, H., Liu, R., Tao, C., Fan, X. and Wang, Z.L., Micro-cable structured textile for simultaneously harvesting solar and mechanical energy. *Nature Energy*, 2016, 16138.)

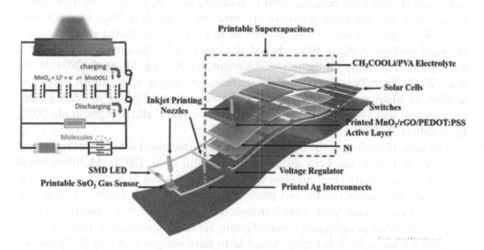

FIGURE 9.3 A wearable self-powered sensor incorporating SnO_2 gas sensor, light-emitting diode (LED), PV supply, supercapacitor storage and voltage regulation on a PET substrate. The schematic indicates that most of the components are printable. The wearable wristband detects methanol or acetone and lights up a warning LED, powered by flexible amorphous silicon solar cells, with energy stored in supercapacitors during standby. (From Lin, Y., Chen, J., Tavakoli, M.M., Gao, Y., Zhu, Y., Zhang, D., Kam, M., He, Z. and Fan, Z., Printable Fabrication of a Fully Integrated and Self-Powered Sensor System on Plastic Substrates, *Adv Mater*, 2019, 31, 1804285.)

Varma et al.'s earlier review (2018) of fibre devices for wearables includes both polymer and perovskite PV as well as DSSC and notes the desirability of biocompatible and breathable materials. They describe Li-ion fibre-shaped batteries that use Ni-coated polyester yarn or predominantly CNT electrodes, having energy densities up to 110 mA h/g whilst being both flexible and stable. Asymmetric supercapacitors with CNT electrodes then had power densities of over 480 W/kg[1] (and energy densities of 42 Wh/kg[1]), but metallic electrodes gave higher values. In contrast to fibre-shaped devices, screen-printed supercapacitors have been put on to cotton and polyester fabrics and on to woven or knitted carbon fibre, but there were still plenty of challenges to making a commercially attractive flexible battery for wearables. A novel combination is described of a perovskite PV that charges an adjacent supercapacitor based on copper ribbon electrodes with MnO_2 filling, woven into a fabric (Li et al., 2016). Electrons generated by illumination through the PV upper electrode (ITO on PET) flow into the supercapacitor via a common PCBM electrode providing energy for operation in the dark. Three tables summarise materials and performances of wearable energy generating and storing devices (Figure 9.4).

9.5 INTEGRATED SYSTEMS WITH SENSORS AND ENERGY PROVISION

The integration of power harvesting with electrical storage in an all-flexible structured device has then been used to drive a wide range of sensors, especially for wearable bioelectronics. These devices exploit both chemical and physical fabrication methods and have been reported to measure motion, strain, pulse rate, sweat composition, ambient gases, humidity, pressure and temperature. They set challenges in stability and biocompatibility, as well as in providing sufficient energy storage in small package systems. Fibre-shaped sensors have limitations at present, often having poor selectivity for the intended measurand (See e.g. Ma et al., 2021; Wang et al., 2020). There is as yet no set of design rules for integrating electronic fibres into textiles. In addition to the sensors and their interconnected power sources, a complete system will also need data storage or transmission (Satharasinghe et al., 2020; Komolafe et al., 2021).

The issue of scaling up the manufacture of such integrated sensing systems based on fibre electronics has been discussed in Seyedin et al. (2021). Multifunctional e-textiles may require electronic and optoelectronic devices as well as sensors (e.g. field-effect transistors (FETs) and photodetectors). Although successive improvements by international teams have introduced a range of materials, mostly organic, to give enhanced performances, manufacture remains complicated, and the products are not low cost. It is highly desirable to have designs that may be fabricated by conventional textile weaving or knitting, and a few examples are given in this paper, mostly of sensors with supercapacitors. They include a terylene glove that triboelectrically sensed the pressure when it grasped different objects via stitched PAN (polyacrylonitrile) yarn pads, and a stretchable textile that contained strain sensor and supercapacitor layers sprayed on to a knitted textile. Wang et al. (2020) also review the challenges in applying fibre-shaped electronics on a commercial

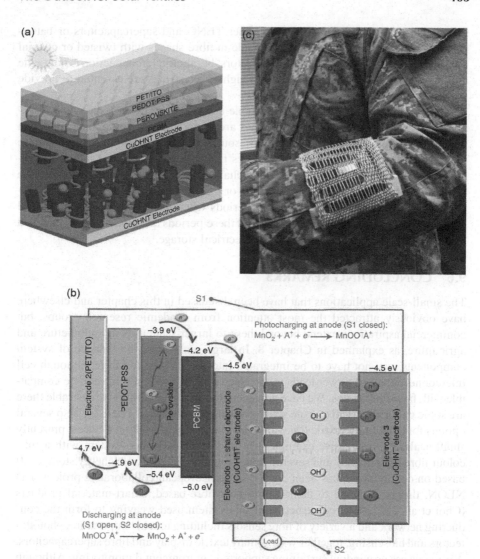

FIGURE 9.4 (a) and (c) Energy harvesting and storing ribbon with perovskite PV on top of a supercapacitor, interconnected by a shared copper electrode that is coated with copper hydroxide nanotubes (CuOHNT). (b) Photogenerated electrons flow into the supercapacitor and are stored at the anode (electrode 1) by an ion reaction; a reverse reaction occurs across the potassium hydroxide gel electrolyte, releasing electrons into the cathode (electrode 3), from which the circuit is externally completed to the PV upper contact (electrode 1). A load may be connected across the supercapacitor. (From Li, C., Islam, M.M., Moore, J., Sleppy, J., Morrison, C., Konstantinov, K., Dou, S.X., Renduchintala, C. and Thomas, J., Wearable energy-smart ribbons for synchronous energy harvest and storage, *Nat. Commun.* 2016, 7, 13319.)

production scale, integrating PV with either TENG and supercapacitors or batteries. The latter storage devices may be made in fibre shapes with twisted or coaxial constructions, especially with CNT conductors, but their interconnection is not facile and may give high resistance losses, although series strings are essential to provide sufficient energy for some systems.

The energy demanded by some wearable e-textiles is still an issue when washability, comfort and other specified wishes are brought in. Energy management can be a useful control when more than one source is being used but at the expense of greater complexity and more electronics that must be integrated into the package. Physiological sensors are particularly challenging and data transmission may be essential, for which there are no standards or material specifications. Each application will have a different profile for the periods when measurements are made and the data are processed and transmitted, and these periods are unlikely to be matched by the energy-harvesting supply without electrical storage.

9.6   CONCLUDING REMARKS

The small-scale applications that have been discussed in this chapter and elsewhere have obviously attracted the most attention from academic research groups, but commercial aspirations extend beyond these to larger-scale uses in architecture and agriculture, as explained in Chapter 8. In large-scale uses, the balance of system components may not have to be integrated into the PV textile fabric, although cell interconnections and power leads to the electrical load will still have to be compatible with fabric substrates. We have now seen that where integration is desirable there are some clever electrical storage solutions that meet the brief. There are also several options for combining textile PV with other electricity harvesting devices, presently small scale. An impressive example of a mid-scale integrated system with a full-colour fibre LED display and several inputs, together with flexible energy storage, all based on cotton fabric, has been created in the European collaborative project, 1-D NEON, that is designed to demonstrate new fibre-based, smart-material products (Choi et al., 2022). The construction of this system used weaving to form the conducting network and a variety of fibre sensors including touch, temperature, photodetectors and biosensing, together with a smart textile display and fibre supercapacitors. A possible use as window curtaining supported environmental monitoring. Although no PV generation was included in this disclosure, it should be realisable by integrating one of the textile PV types discussed in this book.

On a larger scale, the case for architects and builders to use textile envelope, flexible photovoltaics for building-integrated power supply has yet to be met by the supply of product. There are also gaps in the standards and specifications for PV membrane materials that add barriers to putting PV envelopes into practice, despite some elegant and impressive buildings around the world in recent years. See for example the review by Li and Zanelli (2021). Nonetheless, these shortcomings are being addressed and the future for textile PV at all scales of usage is upbeat. We are optimistic that products, from smart clothing to disaster-response shelter to architectural facades, will become more widely available as the world moves towards smarter energy supplies that do not impair the environment.

REFERENCES

Chen, J., Huang, Y., Zhang N., Zou, H., Liu, R., Tao, C., Fan, X. and Wang, Z.L. (2016) Micro-cable structured textile for simultaneously harvesting solar and mechanical energy. *Nat. Energy* 1, 16138 (8 pp.).

Choi, H.W., Shin, D., Yang, J., Lee, S., Figueiredo, C., Sinopoli, S., Ullrich, K., Jovancic, P., Marrani, A., Momente, R., Gomes, J., Branquinho, R., Emanuele, U., Lee, H., Bang, S.Y., Jung, S., Han, S.D., Zhan, S., Harden-Chaters, W., Suh, Y., Fan, X., Lee, T.H., Chowdury, M., Choi, Y., Nicotera, S., Torchia, A., Mincunill, F.M., Candel, V.G., Duraes, N., Chang, K., Cho, S., Kim, C., Lucassen, M., Nejim, A., Jimenez, D., Springer, M., Lee, Y., Cha, S., Sohn, J.I., Igreja, R., Song, K., Barquinha, P., Martins, R., Amaratunga, G.A.J., Occhipinti, L.G., Chhowalla, M. and Kim, J.M. (2022) Smart textile lighting/display system with multifunctional fibre devices for large scale smart home and IoT applications. *Nat. Commun.* 13, 814 (10 pp).

Devadiga, D. et al. (2022). The integration of flexible dye-sensitized solar cells and storage devices towards wearable self-charging power systems: A review. *Renew. Sustain. Energy Rev.* 159, 112252 (35 pp).

Dong, K. and Wang, Z.L. (2021). Self-charging power textiles integrating energy harvesting triboelectric nanogenerators with energy storage batteries/supercapacitors. *J. Semiconductors* 42, 101601 (15 pp.).

Gao, Z., Bumgardner, C., Song, N., Zhang, Y., Li, J. and Li, X. (2016). Cotton-textile-enabled flexible self-sustaining power packs via roll-to-roll fabrication. *Nat. Commun.*, 7, 11586 (12 pp.).

Jiang, Y., He, A., Zhao, R., Chen, Y., Liu, G., Lu, H., Zhang, J., Zhang, Q., Wang, Z., Zhao, C., Long, M., Hu, W., Wang, L., Qi, Y., Gao, J., Wu, Q., Ge, X., Ning, J., Wee, A.T.S. and Qiu, C. (2021). Coexistence of photoelectric conversion and storage in van der Waals heterojunctions. *Phys. Rev. Lett.*, 127, 217401.

Komolafe, A., Zaghari, B., Torah, R., Weddell, A.S., Khanbareh, H., Tsikriteas, Z.M., Vousden, M., Wagih, M., Jurado, U.T., Shi, J., Yong, S., Arumugam, S., Li, Y., Yang, K., Savelli, G. White, N.M. and Beeby, S. (2021). E-textile technology review–From materials to application. *IEEE Access* 9, 97152–97179.

Li, C., Islam, M.M., Moore, J., Sleppy, J., Morrison, C., Konstantinov, K., Dou, S.X., Renduchintala, C. and Thomas, J. (2016). Wearable energy-smart ribbons for synchronous energy harvest and storage. *Nat. Commun.*, 7, 13319.

Li, Q. and Zanelli, A. (2021). A review on fabrication and applications of textile envelope integrated flexible photovoltaic systems. *Renew. Sustain. Energy Rev.*, 139, 110678 (17 pp.).

Ma, X., Jiang, Z. and Lin, Y. (2021). Flexible energy storage devices for wearable bioelectronics. *J. Semiconductors* 42, 101602 (14 pp.).

Ostfeld, A.E., Gaikwad, A.M., Khan, Y. and Arias A.C. (2016). High-performance flexible energy storage and harvesting system for wearable electronics. *Sci. Rep.*, 6, 26122 (10 pp.).

Satharasinghe, A. Hughes-Riley, T. and Dias, T. (2020). A review of solar energy harvesting electronic textiles. *Sensors* 20, 5938 (39 pp.).

Seyedin, S., Carey, T., Arbab, A., Eskandarian, L., Bohm, S., Kim, J.M. and Torrisi, F. (2021). Fibre electronics: Towards scaled-up manufacturing of integrated e-textile systems. *Nanoscale* 13, 12818 (30 pp.).

Sun, H., Zhang, Y., Zhang, J., Sun, X. and Peng, H. (2017). Energy harvesting and storage in 1D devices. *Nat. Rev. Mater.*, 2, 17023.

Varma, S.J., Kumar, K.S., Seal, S., Rajaraman, S. and Thomas, J. (2018). Fiber-type solar cells, nanogenerators, batteries, and supercapacitors for wearable applications. *Adv. Sci.*, 5, 1800340 (32 pp.).

Wang, L., Fu, X., He, J., Shi, X., Chen, T., Chen, P., Wang, B. and Peng, H. (2020). Application challenges in fiber and textile electronics. *Adv. Mater.*, 32, 1901971 (25 pp).

Wen, Z., Yeh, M., Guo, H., Wang, J., Zi, Y., Xu, W., Deng, J., Zhu, L., Wang, X., Hu, C., Zhu, L., Sun, X. and Wang, Z.L. (2016). Self-powered textile for wearable electronics by hybridizing fiber-shaped nanogenerators, solar cells, and supercapacitors. *Sci. Adv.*, 2, 1600097 (9 pp.).

Zhai, S., Karahan, H.E., Wei, L., Qian, Q., Harris, A.T., Minett, A.I., Ramakrishna, S., Ng, A.K. and Che, Y. (2016). Textile energy storage: Structural design concepts, material selection and future perspectives. *Energy Storage Mater.*, 3, 123–139.

Zhao, J., Zha, J., Zeng, Z. and Tan, C. (2021). Recent advances in wearable self-powered energy systems based on flexible energy storage devices integrated with flexible solar cells. *J. Mater. Chem. A.*, 9, 18887 (20 pp.).

Index

Printed in the United States
by Baker & Taylor Publisher Services

Printed in the United States
by Baker & Taylor Publisher Services